The Female Pelvis - Anatomy, Function and Disorders

Edited by Ran Pang,
Jianxin Lu and Xinyao Zhou

Published in London, United Kingdom

The Female Pelvis - Anatomy, Function and Disorders
http://dx.doi.org/intechopen.102196
Edited by Ran Pang, Jianxin Lu and Xinyao Zhou

Contributors
Akanksha Tripathi, Jacob Rubin, Nicole Fleischmann, Novera G. Chughtai, Ran Pang, Samia Aijaz, Shikha Sharma, Summera Malik, Teresa L. Danforth, Urooj Kashif, Yueying Chen

First published in London, United Kingdom, 2024 by IntechOpen
IntechOpen is the global imprint of INTECHOPEN LIMITED, registered in England and Wales, registration number: 11086078, 167-169 Great Portland Street, London, W1W 5PF, United Kingdom

British Library Cataloguing-in-Publication Data
A catalogue record for this book is available from the British Library

Additional hard and PDF copies can be obtained from orders@intechopen.com

The Female Pelvis - Anatomy, Function and Disorders
Edited by Ran Pang, Jianxin Lu and Xinyao Zhou
p. cm.
Print ISBN 978-1-83969-656-5
Online ISBN 978-1-83969-657-2
eBook (PDF) ISBN 978-1-83969-658-9

For EU product safety concerns:
IN TECH d.o.o., Prolaz Marije Krucifikse Kozulić 3, 51000 Rijeka, Croatia,
info@intechopen.com or visit our website at intechopen.com.

Meet the editors

Ran Pang is a consultant urologist and leader in functional urology and urodynamics at Guang'anmen Hospital, China Academy of Chinese Medical Sciences (CACMS). After completing residency training, he was accepted to a clinical fellowship with Peking University, China, in 2005. Subsequently, he joined a research fellowship at Mayo Clinic, USA, in 2011 and a urodynamic fellowship at Dalhousie University, Canada, in 2015. As a leading expert, Prof. Pang also serves on several international organizations, as well as local professional committees. He is the chair of the Publication and Communication Committee, International Continence Society; vice chair of the Kidney Cancer Committee of Integrated Chinese and Western Medicine, China Anti-Cancer Association; and vice chair of the Pelvic Floor Disorder Group of the Urology Committee, Chinese Association of Integrative Medicine.

Jianxin Lu is a director in the Urology Department of Guang'anmen Hospital, China Academy of Chinese Medical Sciences (CACMS). After completing residency training, he was accepted to a clinical fellowship with Peking University, China, in 2001. Later, he participated in the first batch of post-doctoral studies of the State Administration of Traditional Chinese Medicine and was awarded one of twenty-eight national postdocs who achieved outstanding grades. In addition, Prof. Lu serves on a number of professional committees. He is the chair of the Urology Committee, Beijing Association of Integrative Medicine; vice chair of the Prostate Cancer Committee of Integrated Chinese and Western Medicine, China Anti-Cancer Association; and vice chair of the Urology Committee, Chinese Association of Integrative Medicine.

Xinyao Zhou is an associate professor in the Rheumatology Department of Guang'anmen Hospital, China Academy of Chinese Medical Sciences (CACMS). She was nominated as an outstanding young innovative scientific and technological talent of CACMS and the inheritor of the academic experience of traditional Chinese medicine experts in Beijing. She serves on several professional committees. She is the deputy general secretary of community of Innovation in Prevention and Treatment of Rheumatic Diseases, China Association of Chinese Medicine; secretary of the Youth Committee of Rheumatology, Chinese Association of Traditional Chinese Medicine; and a member of the Rheumatology and Immunology Committee, Women's Physician Association. She has been awarded four national invention patents and is the recipient of the First Prize of Scientific and Technological Award, Chinese Society of Traditional Chinese Medicine.

Contents

Preface

With its complex structure, the female pelvis is a vital component of the body. Generally, the female pelvis is broader and wider than the male pelvis because it needs to allow a baby to be delivered. More importantly, the female pelvis provides powerful support for the bladder, urethra, productive organs, and rectum. In Chapter 1, Dr. Ran Pang shows an overview of the female pelvis.

In terms of female pelvic anatomy, the perineum plays a critical role in maintaining normal pelvic function. Anatomically, the perineum can be divided into two triangular regions: the urogenital triangle anteriorly and the anal triangle posteriorly. The former acts as a guard and sling for the urethra and vagina, while the latter provides support and protection for the anal canal. In Chapter 2, Dr. Novera G. Chughtai et al. explain the intricate anatomy and embryology of the perineum in an easy-to-understand manner.

Compared to men, women's pelvises are more susceptible to injury due to several special conditions, such as pregnancy and childbirth. Besides, several factors, including low estrogen status, increasing age, obesity, chronic constipation, and diabetes, may weaken or damage the tension of the pelvic muscles and connective tissues and consequently result in pelvic floor dysfunction (PFD). In addition to outlining PFDs, Dr. Yueying Chen and Ran Pang, in Chapter 3, concentrates on two of the most prevalent PFDs: urine incontinence and pelvic organ prolapse.

Urinary fistulas, as a kind of complex PFD, can be divided into vesicovaginal fistula, ureterovaginal fistula, urethrovaginal fistula, vesicouterine fistula, colovesical fistula, and ureteroarterial fistula, depending on the fistula location. Surgical techniques have evolved enormously over the past few decades. In Chapter 4, Dr. Jacob Rubin and Dr. Teresa L. Danforth discuss the advantages and disadvantages of different surgical approaches for each urinary fistula.

Endometriosis is an estrogen-dependent, inflammatory disease that is most commonly located in the pelvis. Despite its benign nature, endometriosis can lead to a number of conditions, such as dysmenorrhea, dyspareunia, chronic pain, and infertility. In Chapter 5, Dr. Shikha Sharma and Akanksha Tripathi provide a thorough overview of endometriosis from a clinical standpoint. Not only are conservative treatments elaborated, but complementary therapies are also introduced.

Behavioral treatments have been the basic therapeutic strategy for PFDs. The prerequisite for these treatments to be effective is pelvic floor awareness and proper education. In Chapter 6, Dr. Nicole Fleischmann presents the role of re-toilet training in the management of PFDs by analyzing the relationship between unhealthy toilet habits in childhood and adult PFDs.

In this book, experts and researchers from different countries present the latest advances and innovations in the field of the female pelvis. These chapters offer readers vital updates, even though they cannot cover every facet of the female pelvis.

Ran Pang and Jianxin Lu
Department of Urology,
Guang'anmen Hospital,
China Academy of Chinese Medical Sciences,
Beijing, China

Xinyao Zhou
Department of Rheumatology,
Guang'anmen Hospital,
China Academy of Chinese Medical Sciences,
Beijing, China

Section 1

Introduction

Chapter 1

Introductory Chapter: A Quick Glance at Female Pelvis

Ran Pang

1. Introduction

Female pelvis plays a critical role in providing protection and support for the bladder, urethra, productive organs, and rectum. Anatomically, three different systems including bone, muscle, and connective tissue act as the guard and sling functions [1]. Of those, the female bony pelvis consists of the sacrum and two hip bones which are formed by the fusion of the ilium, ischium, and pubis. The bony pelvis provides not only protection for pelvic organs but also an essential framework for the muscular and fascial supports. The levator ani muscles, which are composed of the pubococcygeus, puborectalis, and iliococcygeus, play an important part in slinging the pelvic organs, since the urethra, vagina, and rectum pass through the levator ani hiatus. The normal levator ani muscle tone is sufficient to maintain the hiatus closed, which serves as a support to pelvic organs. A voluntary contraction of levator ani muscles can make the hiatus further closed by compassing mid-urethra, vagina, and rectum against the pubic bone, which contributes to resisting excess load. The pelvic connective tissue composed of endopelvic fasciae and ligaments is also responsible for holding the pelvic organs in place. In addition, the coordination between levator ani muscles and pelvic floor fasciae may further strengthen the pelvic floor support. As is known, the connective tissue in human beings has adoptive extensibility which means it may be stretched by subjecting to continuous stress. However, the adoptive extension is not observed in the fasciae and ligaments within pelvis because the resting tone of levator ani muscles bear the weight of abdominal and pelvic organs [2]. The normal function of these three systems guarantees female pelvic health.

In general, the female pelvis is more delicate than the male pelvis since both pregnancy and vaginal delivery may lead to a significant increase in the possibility of pelvic floor dysfunction (PFD). According to the terminology released by International Urogynecological Association (IUGA) and International Continence Society (ICS), PFD is an umbrella diagnosis covering a number of disorders including bladder and bowel dysfunction, pelvic organ prolapse (POP), sexual dysfunction, and pelvic pain [3]. Based on two epidemiological studies, as high as 46.5% to 67.5% of women suffered from at least one type of PFDs [4, 5]. Of those women, approximately one-third experienced two or more types of PFDs and one-fifth failed conservative management and required surgical intervention [4, 6, 7]. Not only does PFD diminish women's quality of life significantly, but its related costs also pose a huge burden for the healthcare system [8, 9]. In terms of specific disorders, urinary incontinence and POP are the most common types of PFDs.

 IntechOpen

2. Urinary incontinence

Urinary incontinence is one of the most common PFDs, with prevalence ranging from 9.3 to 69%, depending on the study population and methodological difference [10]. Based on different manifestations, urinary incontinence can be classified into three subtypes: stress urinary incontinence (SUI), urgency urinary incontinence (UUI), and mixed urinary incontinence (MUI). ICS defines SUI as involuntary urine leakage with effort or physical exertion, UUI as involuntary urine leakage accompanied by urgency, and MUI as the coexistence of both SUI and UUI [11].

Urodynamic study provides theoretical insights in normal continence mechanism and the pathogenesis of urinary incontinence subtypes. In order to maintain continence, the urethral pressure must be higher than the bladder pressure, both at rest and during activities. It is traditionally believed that urethral striated muscles (urethral sphincter), smooth muscle, and vascular bed within the submucosa contribute equally to urethral pressure at rest [12]. However, more recent studies demonstrated that the spontaneous activity of circular and longitudinal urethral smooth muscles provided resting urethral pressure, while urethral striated muscles were not involved [13, 14]. By contrast, the urethral striated muscles play an important role in providing additional urethral pressure under conditions of physical stress. During physical activity, the contraction of the diaphragm and abdominal muscles can result in a sudden increase in abdominal pressure, which is transmitted to bladder, causing a corresponding increase in bladder pressure. It is observed that a hard cough can lead to an increase in bladder pressure by more than 200 cm H_2O [2]. In response to the stress, urethral striated muscles contract reflexively, causing an elevation in urethral pressure. Besides, the reflex contraction of the levator ani muscles can also enhance urethral pressure by tensing the sub-urethral fasciae, which decreases vesical neck motion [15, 16]. The damage or dysfunction of levator ani muscles and/or urethral striated muscles can result in SUI. It is widely recognized that urethral hypermobility and intrinsic sphincter deficiency (ISD) are the exact etiology of SUI [17]. Impaired anatomical support of urethra due to levator ani muscles dysfunction is the main reason of urethral hypermobility [18]. Some studies, using translabial ultrasound or magnetic resonance imaging (MRI), showed that the area and circumference of urethral striated muscles were shorter and smaller in women with SUI in comparison with the controllers [19, 20]. Compared to SUI, only limited studies reported the association between pelvic anatomical abnormalities and UUI. Wang et al. found that women with overactive bladder (OAB) or UUI had weaker urethral sphincter and levator ani muscles, causing an over-relaxed levator ani hiatus compared to health controllers [21]. The more understanding of the pathogenesis of urinary incontinence we have, the more precise treatment strategy we can develop.

In terms of the management of urinary incontinence, lifestyle modification and behavioral therapy have been considered as the foundation of treatment, no matter what subtype the patient has [22]. Pelvic floor muscle training (PFMT), as the first-line treatment of SUI, can strengthen levator ani muscles and consequently enhance urethral support [23]. The effectiveness of PFMT for SUI has been confirmed by moderate to high certainty evidence [24]. However, adherence to PFMT has been problematic. It is reported that the short-term and long-term adherence to PFMT is only 64% and 23%, respectively [25]. With the increasing use of social media, a number of mobile applications have been developed as an assisted tool of PFMT. A study showed that mobile applications may improve the potential adherence to PFMT and enhance the effect of treatment [26]. Initially, PFMT was designed for

SUI treatment, and some studies further revealed it could be an effective approach for OAB and UUI. Evidence showed that the contraction of pelvic floor muscles could suppress the detrusor overactivity and relieve the urgency symptoms [27]. For patients with urinary incontinence failed to behavioral therapy, pharmacotherapy can be considered. Although a Cochrane systematic review demonstrated that duloxetine, a serotonin-noradrenaline reuptake inhibitor, could improve SUI symptoms significantly [28], it has not been approved by U.S. Food and Drug Administration (FDA) for the treatment of SUI. Antimuscarinics and beta-agonists are the two different kinds of medications used for the treatment of OAB and UUI. The former can suppress detrusor overactivity by inhibiting parasympathetic nervous system, while the latter can relax detrusor by activating sympathetic nervous system. For patients who do not respond to pharmacotherapy, minimally invasive and surgical intervention can be the treatment options. The mid-urethral sling, pubovaginal sling, and urethral bulking agents are the common procedures for SUI [22], whereas sacral neuromodulation, percutaneous tibial nerve stimulation, and intradetrusor injection of onabotulinumtoxin A are used for refractory OAB or UUI [29]. In addition, growing evidence demonstrates acupuncture may act as an optional treatment for SUI [30].

3. Pop

Pelvic organ support depends on the normal structure and function of pelvic floor muscles and connective tissues. The impairment of levator ani muscles and/or connective tissues may result in POP. A number of studies presented a strong association between levator ani defects and POP [31]. It is reported that 55% of women with POP had major levator ani defects, and 83% of women with levator ani defects presented POP [32, 33]. Furthermore, two important theories of pelvic organ support demonstrated the role of pelvic connective tissues in the development of POP.

In 1990, Petros et al. proposed their "Integral Theory" [34]. According to the theory, the pelvic organs are supported in their normal locations by a group of pelvic floor fasciae and ligaments. Functionally, these fasciae and ligaments act like suspension bridges, and their over-laxity can lead to POP and related symptoms. The integral theory further divided the female pelvis into three areas: anterior, middle, and posterior, which are used to indicate the impaired area and implement precision treatment [35].

DeLancey described his anatomic "hammock theory" in 1992, which divided the support of the pelvic organs into three levels [36]. Level 1 support refers to uterosacral ligament's suspension to the cervix and upper vagina. Failure of level I support results in the uterine or vaginal vault prolapse (apical prolapse). Level 2 support consists of the vagina's lateral connections to the arcus tendineus fasciae pelvis, which can hold up the mid-vagina. Failure of level II support leads to anterior and posterior vaginal wall prolapse, as well as SUI. Level 3 support is comprised of the connections between the perineal membrane and perineal body fasciae, which can support the lower one-third of vagina. Failure of level II support results in SUI or distal rectocele [37, 38].

The reported prevalence of POP varies from 3 to 50% depending on the different study population and investigative methodology [39]. The confirmed risk factors for POP mainly include age, obesity, increasing parity, vaginal delivery, connective tissue disorders, menopausal status, and chronic constipation [40]. The management of POP is based on patients' symptoms, degree of prolapse, and influence on daily life. In general, observation is recommended for women with asymptomatic POP [40].

The conservative therapy including pelvic floor physical therapy, biofeedback, and electrogalvanic stimulation can be considered as the initial treatment for symptomatic POP. Moreover, pessary is also an optional therapy with a reported success rate of 93% [41]. For women with POP who failed conservative treatment, surgical intervention can be an effective treatment option. Surgery can be divided into reconstructive and obliterative procedures. The former focuses on maintaining the vaginal canal, whereas the latter centers on closing the vaginal canal.

4. Summary

Female pelvis is a complex anatomical region composed of bony pelvis, pelvic muscle, connective tissue, and pelvic organs. The levator ani muscles as well as pelvic floor fasciae and ligaments play a critical role in maintaining pelvic organs in normal location. The injury and defects of these muscles and connective tissues can result in PFDs. With a deep understanding of functional anatomy of female pelvis, precision treatment can be achieved for the management of PDFs.

Acknowledgements

This work was supported by the National Natural Science Foundation of China (81804069), the Science and Technology Innovation Project of China Academy of Chinese Medical Sciences (CI2021A01502), and grant 2019S443 from the Guang'anmen Hospital, China Academy of Chinese Medical Sciences.

Conflict of interest

None.

Author details

Ran Pang
Guang'anmen Hospital, China Academy of Chinese Medical Sciences, Beijing, China

*Address all correspondence to: pangran2002@gmail.com

References

[1] Tamilselvi A, Rane A. Principles and Practice of Urogynaecology. New Delhi, India: Springer; 2015

[2] Ashton-Miller JA, DeLancey JO. Functional anatomy of the female pelvic floor. Annals of the New York Academy of Sciences. 2007;**1101**:266-296

[3] Haylen BT, de Ridder D, Freeman RM, Swift SE, Berghmans B, Lee J, et al. An international Urogynecological association (IUGA)/international continence society (ICS) joint report on the terminology for female pelvic floor dysfunction. International Urogynecology Journal. 2010;**21**(1):5-26

[4] Gyhagen M, Åkervall S, Milsom I. Clustering of pelvic floor disorders 20 years after one vaginal or one cesarean birth. International Urogynecology Journal. 2015;**26**(8):1115-1121

[5] Kepenekci I, Keskinkilic B, Akinsu F, Cakir P, Elhan AH, Erkek AB, et al. Prevalence of pelvic floor disorders in the female population and the impact of age, mode of delivery, and parity. Diseases of the Colon and Rectum. 2011;**54**(1):85-94

[6] Nygaard I, Barber MD, Burgio KL, Kenton K, Meikle S, Schaffer J, et al. Prevalence of symptomatic pelvic floor disorders in US women. Journal of the American Medical Association. 2008;**300**(11):1311-1316

[7] Wu JM, Matthews CA, Conover MM, Pate V, Jonsson FM. Lifetime risk of stress urinary incontinence or pelvic organ prolapse surgery. Obstetrics and Gynecology. 2014;**123**(6):1201-1206

[8] Davis KJ, Kumar D, Wake MC. Pelvic floor dysfunction: A scoping study exploring current service provision in the UK, interprofessional collaboration and future management priorities. International Journal of Clinical Practice. 2010;**64**(12):1661-1670

[9] Van der Woude DA, Pijnenborg JM, de Vries J. Health status and quality of life in postpartum women: A systematic review of associated factors. European Journal of Obstetrics, Gynecology, and Reproductive Biology. 2015;**185**:45-52

[10] Milsom I, Gyhagen M. The prevalence of urinary incontinence. Climacteric. 2019;**22**(3):217-222

[11] Abrams P, Cardozo L, Fall M, Griffiths D, Rosier P, Ulmsten U, et al. The standardisation of terminology in lower urinary tract function: Report from the standardisation sub-committee of the international continence society. Urology. 2003;**61**(1):37-49

[12] Rud T, Andersson KE, Asmussen M, Hunting A, Ulmsten U. Factors maintaining the intraurethral pressure in women. Investigative Urology. 1980;**17**(4):343-347

[13] Sergeant GP, Hollywood MA, Thornbury KD. Spontaneous activity in urethral smooth muscle. Advances in Experimental Medicine and Biology. 2019;**1124**:149-167

[14] Venema PL, Kramer G, van Koeveringe GA, Heesakkers J. The maximal urethral pressure at rest and during Normal bladder filling is only determined by the activity of the urethral smooth musculature in the female. Journal of Clinical Medicine. 2023;**12**(7):2575

[15] Dietz HP, Erdmann M, Shek KL. Reflex contraction of the levator ani in

women symptomatic for pelvic floor disorders. Ultrasound in Obstetrics & Gynecology. 2012;**40**(2):215-218

[16] Miller JM, Perucchini D, Carchidi LT, DeLancey JO, Ashton-Miller J. Pelvic floor muscle contraction during a cough and decreased vesical neck mobility. Obstetrics and Gynecology. 2001;**97**(2):255-260

[17] Kalejaiye O, Vij M, Drake MJ. Classification of stress urinary incontinence. World Journal of Urology. 2015;**33**(9):1215-1220

[18] Falah-Hassani K, Reeves J, Shiri R, Hickling D, McLean L. The pathophysiology of stress urinary incontinence: A systematic review and meta-analysis. International Urogynecology Journal. 2021;**32**(3): 501-552

[19] Cassadó Garriga J, Pessarrodona Isern A, Rodríguez Carballeira M, Pallarols Badia M, Moya Del Corral M, Valls Esteve M, et al. Three-dimensional translabial ultrasound assessment of urethral supports and the urethral sphincter complex in stress urinary incontinence. Neurourology and Urodynamics. 2017;**36**(7):1839-1845

[20] Kim JK, Kim YJ, Choo MS, Cho KS. The urethra and its supporting structures in women with stress urinary incontinence: MR imaging using an endovaginal coil. AJR. American Journal of Roentgenology. 2003;**180**(4):1037-1044

[21] Wang Y, Yao J, Chen N, Liu J, Shi G, Wu Y. Study of female pelvic floor muscle in overactive bladder based on MRI 3D reconstruction. BMC Urology. 2022;**22**(1):132

[22] Vaughan CP, Markland AD. Urinary Incontinence in women. Annals of Internal Medicine. 2020;**172**(3):Itc17-itc32

[23] Sheng Y, Carpenter JS, Ashton-Miller JA, Miller JM. Mechanisms of pelvic floor muscle training for managing urinary incontinence in women: A scoping review. BMC Women's Health. 2022;**22**(1):161

[24] Todhunter-Brown A, Hazelton C, Campbell P, Elders A, Hagen S, McClurg D. Conservative interventions for treating urinary incontinence in women: An overview of Cochrane systematic reviews. Cochrane Database of Systematic Reviews. 2022;**9**(9):Cd012337

[25] Dumoulin C, Hay-Smith J, Frawley H, McClurg D, Alewijnse D, Bo K, et al. 2014 consensus statement on improving pelvic floor muscle training adherence: International continence society 2011 state-of-the-science seminar. Neurourology and Urodynamics. 2015;**34**(7):600-605

[26] Latorre GFS, de Fraga R, Seleme MR, Mueller CV, Berghmans B. An ideal e-health system for pelvic floor muscle training adherence: Systematic review. Neurourology and Urodynamics. 2019;**38**(1):63-80

[27] Burgio KL, Locher JL, Goode PS, Hardin JM, McDowell BJ, Dombrowski M, et al. Behavioral vs drug treatment for urge urinary incontinence in older women: A randomized controlled trial. Journal of the American Medical Association. 1998;**280**(23):1995-2000

[28] Mariappan P, Ballantyne Z, N'Dow JM, Alhasso AA. Serotonin and noradrenaline reuptake inhibitors (SNRI) for stress urinary incontinence in adults. Cochrane Database of Systematic Reviews. 2005;**20**(3):Cd004742

[29] Raju R, Linder BJ. Evaluation and treatment of overactive bladder in women. Mayo Clinic Proceedings. 2020;**95**(2):370-377

[30] Cardozo L, Rovner E, Wagg A, Wein A. Incontinence 7th ed., Bristol, UK: International Continence Society; 2023

[31] Notten KJB, Vergeldt TFM, van Kuijk SMJ, Weemhoff M, Roovers JWR. Diagnostic accuracy and clinical implications of Translabial ultrasound for the assessment of Levator Ani defects and Levator Ani biometry in women with pelvic organ prolapse: A systematic review. Female Pelvic Medicine & Reconstructive Surgery. 2017;**23**(6):420-428

[32] DeLancey JO, Morgan DM, Fenner DE, Kearney R, Guire K, Miller JM, et al. Comparison of levator ani muscle defects and function in women with and without pelvic organ prolapse. Obstetrics and Gynecology. 2007;**109**(2 Pt 1):295-302

[33] Dietz HP, Simpson JM. Levator trauma is associated with pelvic organ prolapse. BJOG : An International Journal of Obstetrics and Gynaecology. 2008;**115**(8):979-984

[34] Petros PE, Ulmsten UI. An integral theory of female urinary incontinence. Experimental and clinical considerations. Acta Obstetricia et Gynecologica Scandinavica. Supplement. 1990;**153**:7-31

[35] Petros PE, Woodman PJ. The integral theory of continence. International Urogynecology Journal and Pelvic Floor Dysfunction. 2008;**19**(1):35-40

[36] DeLancey JO. Anatomic aspects of vaginal eversion after hysterectomy. American Journal of Obstetrics and Gynecology. 1992;**166**(6 Pt 1):1717-1724. discussion 24-8

[37] DeLancey JO. The anatomy of the pelvic floor. Current Opinion in Obstetrics & Gynecology. 1994;**6**(4):313-316

[38] Luo J, Betschart C, Chen L, Ashton-Miller JA, DeLancey JO. Using stress MRI to analyze the 3D changes in apical ligament geometry from rest to maximal Valsalva: A pilot study. International Urogynecology Journal. 2014;**25**(2):197-203

[39] Weintraub AY, Glinter H, Marcus-Braun N. Narrative review of the epidemiology, diagnosis and pathophysiology of pelvic organ prolapse. International Braz J Urol. 2020;**46**(1):5-14

[40] American College of Obstetricians and Gynecologists and the American Urogynecologic Society. Pelvic organ prolapse. Female Pelvic. 2019;**25**(6):397-408

[41] Cundiff GW, Amundsen CL, Bent AE, Coates KW, Schaffer JI, Strohbehn K, et al. The PESSRI study: Symptom relief outcomes of a randomized crossover trial of the ring and Gellhorn pessaries. American Journal of Obstetrics and Gynecology. 2007;**196**(4):405.e1-405.e8

Section 2

An Insight into the Anatomy

Chapter 2

Functional Anatomy of Female Perineum

Novera G. Chughtai, Urooj Kashif, Samia Aijaz and Sumera Malik

Abstract

Female perineum is the tissue complex between the peritoneum and the skin that closes the pelvis inferiorly and its functionality depends on the interplay between organs, tissues, septae and spaces in it. It is a diamond-shaped region below the pelvic floor and extends between the pelvic diaphragm and the perineal skin. It is a surprisingly dynamic field with new insights, discoveries, and controversies and carries differences in viewpoint among anatomists and surgeons. This book chapter will provide an overview regarding perineal anatomy in the female and will focus on embryology, anatomy of the perineal region with modern proponents. It includes detailed anatomy of Urogenital and Anal triangles, their muscles with blood supply and innervation, anatomy and functions of the perineum, its role in Pelvic Organ Prolapse and clinical significance in urinary and fecal incontinence and contribution towards common obstetric and gynecological pathologies.

Keywords: female perineum, anatomy, UV prolapse, incontinence, perineal body, urogenital triangle, anal triangle

1. Introduction

The perineum is situated below the pelvic diaphragm and is a diamond-shaped space. It is divided by an imaginary line between the anterior border of the ischial tuberosity into two triangular divisions, anteriorly the urogenital triangle and anal triangle posteriorly [1]. The urinogenital triangle is further divided by a fascial layer called the perineal membrane, into a deep perineal pouch, present above the perineal membrane and superficial perineal space present beneath it. Contents of the superficial perineal space are erectile tissues, skeletal muscles, and terminal branches of the internal pudendal vessels and nerves. Knowledge of the anatomy of the perineum is necessary to provide the best surgical outcomes in cases related to malignancies, trauma and congenital abnormalitiess.

2. Boundaries of the perineal region

The surface projection of the perineum is best visualized in the lithotomy position, with the individual lying in a supine position and the hip joints in partial flexion and

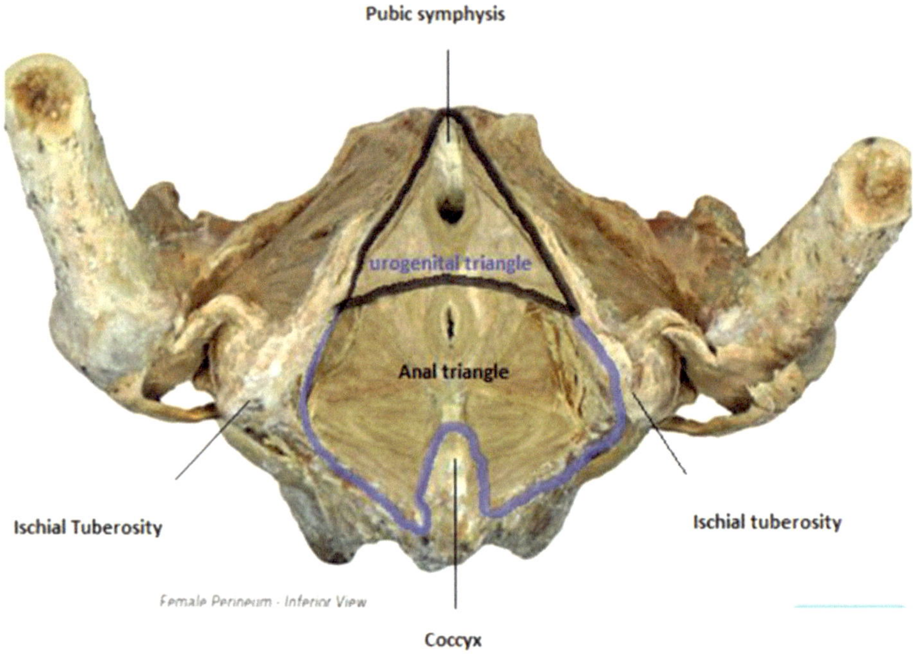

Figure 1.
Boundaries of female perineum and division into urogenital and anal triangles.

abduction. Anteriorly, it is bounded by the arcuate ligament and lower border of the pubic symphysis, posteriorly by the tip of coccyx and ischiopubic ramus, anterolaterally by ischial tuberosities and the Sacro tuberous ligaments posterolaterally [2]. An imaginary inter-ischial line between the anterior ends of the ischial tuberosity divides the perineum into an anterior urogenital triangle and a posterior anal triangle (**Figure 1**).

The perineum is divided into superficial and deep compartments. Which are separated from each other Perineal membrane which is a fibrous connective tissue layer. The perineal membrane spans over the anterior half of the pelvic outlet. It was initially called the urogenital diaphragm; this modification of its name verifies that it is not a two-layered structure with muscle in between, as previously thought. The vagina and urethra are supported by the perineal membrane as they pass through it. Cephalad to the perineal membrane lies the striated urogenital sphincter muscle, which allows compression of the mid- and distal urethra [3, 4].

3. Embryology

Embryologically, the muscular pelvic part, is formed by two distinct muscle groups: the pubocaudal muscle and the muscle of Gegenbauer close to the cloaca.

The Pubocaudal muscle group gives rise to the levator ani and ischio-coccygeal muscle, pubourethral-bladder-uterus-rectosacral ligament and Gegenbauer group, after the descent of the genito-urinary septum forms the urethral and anal sphincter muscles and the superficial and deep transverse muscles, bulbocavernosus, and ischiocavernosus [5]. In fetus, these structures are present as a unit, synergistic in their functions.

The three divisions of the primitive gut are foregut, midgut, and hindgut and are derived from endoderm. The allantois and the hindgut enter the cloaca which is a common cavity and is in contact with the ectodermal surface. An ectodermal depression sinks in towards the gut until only the thin cloacal membrane remains between the gut and the outside. This is called the proctodaeum [6]. Partitioning of the cloacal membrane takes place during the fifth to seventh week of development when the mesodermal urorectal septum descends between the hindgut and the allantois to fuse with the cloacal membrane [7]. The area of fusion becomes the perineal body and separates the dorsal anal membrane from the larger ventral urogenital membrane. By the eighth week, the anal membrane breaks down, establishing the anal canal. The urorectal septum also divides the cloacal musculature into anterior and posterior components and thus explains why the nerve supply of the whole musculature is supplied by one nerve, the pudendal nerve [5].

4. Urogenital triangle

The urogenital triangle corresponds to the anterior region of the perineum. The perineal membrane divides the urogenital triangle into the deep perineal space (located above the membrane) and the superficial perineal space (located below the membrane). The perineal membrane is continuous with the superior fascia of the urogenital diaphragm [8]. At its apical end, the perineal membrane thickens to form the transverse perineal ligament [9]. The perineal membrane is anchored to the inferior surface of the ischio-pubic ramus on each side and extends posteriorly to attach to the perineal body posteriorly.

Through dissection and MRI, the perineal membrane is found to be neither a simple membrane nor a three laminar fascia-muscle-fascia unit. It is a complex multi-structure tissue integrated into a larger interconnected system, with varying histological composition according to the different structures. There is clear evidence of the presence of distinctive anterior and posterior portions that might play different roles [10].

Three fascial layers divide the superficial perineal space further into the subcutaneous perineal pouch and the superficial perineal pouch: the superficial fascia of the perineum, the deep perineal fascia, and the perineal membrane. The subcutaneous perineal pouch is limited by the superficial fascia of the perineum (Colle's fascia) externally and the deep perineal fascia internally. Its thin layer of subcutaneous tissue, provides mobility to the skin over the external genitalia. The superficial perineal pouch is present between the perineal membrane and the deep perineal fascia. It is bounded by the inner surface of the ischiopubic ramus on either side and communicates anteriorly with the potential space between the Scarpa's fascia and the fascia of the abdominal wall muscles [11].

The superficial perineal space contains the ischiocavernosus, bulbospongiosus, superficial transverse perineal muscles (superficial perineal muscles), the erectile tissue of the clitoris, the vestibular bulbs, and Bartholin glands. The deep perineal space lies inferior to the levator ani muscles just deep to the perineal membrane.

Within the deep perineal space lie the external urethral sphincter and the urethrovaginalis, compressor urethrae, and deep transverse perineal muscles. The urethrovaginalis and compressor urethrae muscles provide accessory sphincter function to the urethra. The urethrovaginalis muscles surround the distal urethra and vagina without passing between them, and therefore act as a sphincter to the vagina as well the distal urethra [3, 12].

The deep transverse perineal muscle, along with its superficial counterpart, serves to stabilize the position of the perineal body and inferior border of the perineal membrane. However, there is some dispute as to whether the deep transverse perineal muscle is present in females.

4.1 Muscles of perineum

Muscles of perineum include superficial transverse perineal, Bulbospongiosus and ischiocavernosus muscles (**Figure 2**).

4.1.1 Superficial transverse perineal muscle

The superficial transverse muscle, which arises from the inner and forepart of the ischial tuberosity and is inserted into the central tendinous part of the perineal body. It attaches to the central tendon of the perineal body along with its counterpart from the opposite side, the EAS from behind, and the bulbospongiosus in the front [8].

4.1.2 Bulbospongiosus muscle

It originates from the inferior surface of the superior pubic rami and the crura of the clitoris. Bulbospongiosus fibers blend with the superficial transverse perineal muscles and the external anal sphincter and inserts on the perineal body [13].

4.1.3 Ischiocavernosus muscle

This muscle is elongated, broader at the middle and is situated on the side of the lateral boundary of the perineum. It arises from the inner surface of the ischial tuberosity, behind the crura clitoris, from the surface the adjacent portions of the ischial ramusand surface of the crus [14].

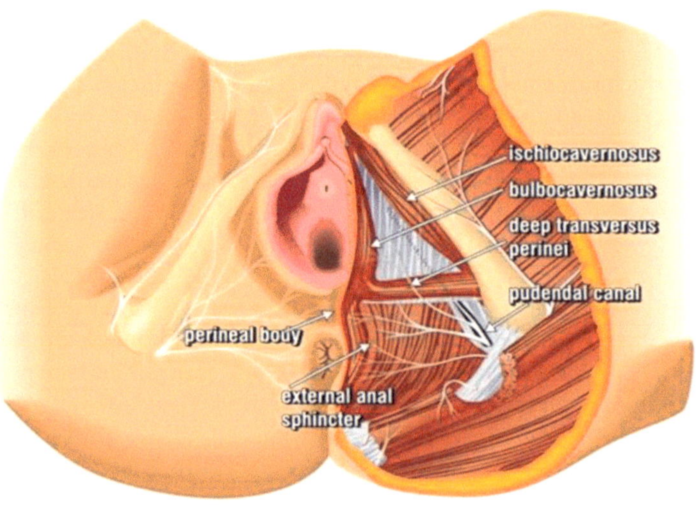

Figure 2.
Muscles of perineum: lower layer of muscles, distal parts of urethra, vagina, rectum and perineal body. Patient supine. View into perineum.

4.2 Urethra

The female urethra is approximately 4 cm in length and 6 mm in diameter. Its lumen perforates the perineal membrane, with its external orifice in the vestibule directly above the vaginal opening. The urethra is embedded in the adventitia of the anterior vagina throughout its length. The urethral epithelium has longitudinal folds and many small glands, which open into the urethra [15]. The epithelium is continuous externally with vulval epithelium and internally with that of the bladder which is stratified squamous and becomes transitional near the bladder. The lamina propria supports the epithelial lining which is a layer of loose fibroelastic connective tissue. It contains collagen fibril bundles and fibrocytes and abundant elastic fibers spread longitudinally and circularly around the urethra. A characteristic feature are numerous thin-walled veins. This rich vascular supply contributes to urethral resistance. The urethral smooth muscle is composed primarily of oblique and longitudinal muscle fibers, with a few circularly oriented outer fibers. The intrinsic urethral sphincter mechanism is constructed by This muscle and the detrusor muscle in the bladder base. This smooth muscle is considered to be under alpha-adrenergic and cholinergic control, although Gosling et al. [16] found an extensive cholinergic nerve supply with few noradrenergic nerves. The longitudinal muscles shorten and widen the urethral lumen during voiding phase and the circular smooth muscle (along with the striated urogenital sphincter muscle) contributes to urethral resistance to outflow at rest. The extrinsic urethral sphincter mechanism is formed by the striated urethral and periurethral muscles. Its inner portion lies within and adjacent to the urethral wall, and outer portion is composed of skeletal muscle fibers of the pelvic diaphragm. The inner portion is made up of the sphincter urethrae, which surrounds the proximal two thirds of the urethra, and the compressor urethrae and urethrovaginal sphincter (known together formerly as the deep transverse perineal muscle), which arch over the ventral surface of the distal one third of the urethra. These three muscles functioning as a single unit, constitute the striated urogenital sphincter [11] which exerts tone on the urethral lumen over prolonged time periods. These muscles with the levator ani also contribute to voluntary interruption of the urine stream and urethral closure with stress via reflex muscle contraction.

4.3 Urethral support

Vagina provides support to the bladder and urethra. Support of the urethra and bladder neck is thought to be provided by the interplay of the pubourethral ligaments, the perineal membrane, and muscles of the pelvic floor.

For decades, the pubourethral and pubovesical ligaments have ben under debate. Few studies describe the pubourethral ligament as being one structure divided into two distinct portions (i.e., superior-inferior) [17] while others mention that it was divided into three portions (i.e., superiorintermediate-inferior), and some found that it comprises of two different ligaments (i.e., pubourethral and pubovesical ligaments).

It is also documented that the pubovesical ligament is an extension of the detrusor muscle and play a role in micturition by assisting vesical neck opening. This contradicts the existence of a so-called "pubovesical ligament" and states that it should be called the pubovesical muscle because its composition is mainly of smooth muscle cells [18].

DeLancey [19–21] reports that rather than being suspended ventrally by ligamentous structures, the proximal urethra and bladder base are supported in a sling model by the anterior vaginal wall, which is attached bilaterally to the levator ani muscles at the arcus tendineus fasciae pelvis (**Figure 3**). These attachments extend caudally and blend with the superior fibers of the perineal membrane. The tissues fix the distal urethra beneath the pubic bone and described as pubourethral ligaments, are composed of the perineal membrane and the lower portion of the arcus tendineus fasciae pelvis. Anterior vaginal attachment at the arcus tendineus fasciae pelvis may contribute to urethral closure by providing a stable base onto which the bladder neck and proximal urethra are compressed with increases in intra-abdominal pressure. These are responsible for the posterior movement of the vesical neck seen at the start of micturition and for the elevation seen when a patient is instructed to hold her urinary stream. Defects in these attachments probably result in anterior vaginal prolapse (cystocele) and proximal urethral support defects (urethral hypermobility) conditions associated with stress urinary incontinence. These defects are also responsible for paravaginal fascial defects, due to vaginal detachment from its lateral connective tissue supports [23].

Coordination of the involuntary detrusor and intrinsic urethral sphincteric muscles ensure proper urinary continence mechanism and may be backed by the voluntary levator ani muscle and external urethral sphincter muscle function. The levator plate resists posterior rotational descent of the vagina and overlying urethrovesical junction during levator contraction providing a dynamic backdrop [24]. Pelvic musculature may be damaged during childbirth; consequently, posterior rotational descent of the bladder neck necessary for normal voiding can also occur at times of increased intra-abdominal pressure.

Relationship between urinary incontinence symptomatology and the number of repeated vaginal deliveries is documented. The risk for urinary incontinence was found to be contributing: 3.3%, 1.0%, and 6.8% after one, two, and three deliveries,

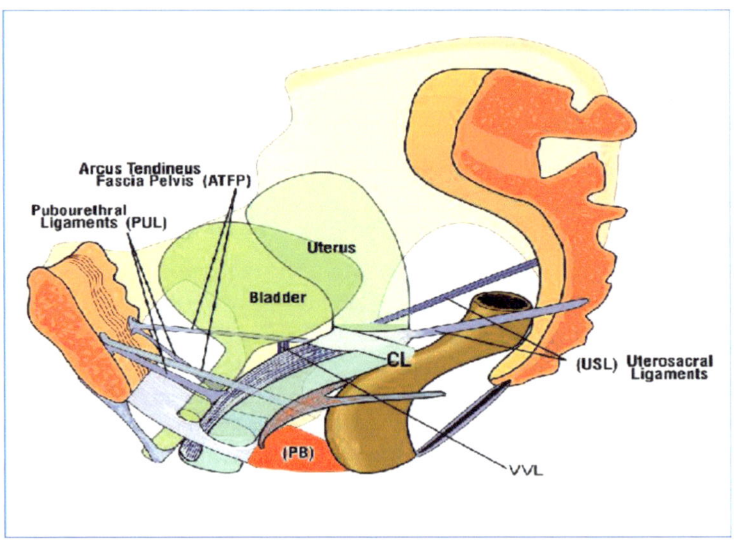

Figure 3.
Pubourethral ligament (PUL), arcus tendineus fascia pelvis (ATFP), cardinal (CL), and uterosacral (USL). The vesico-vaginal ligament (VVL) is the attachment between the bladder base and proximal vagina. With permission from Dr. Peter Petros [22].

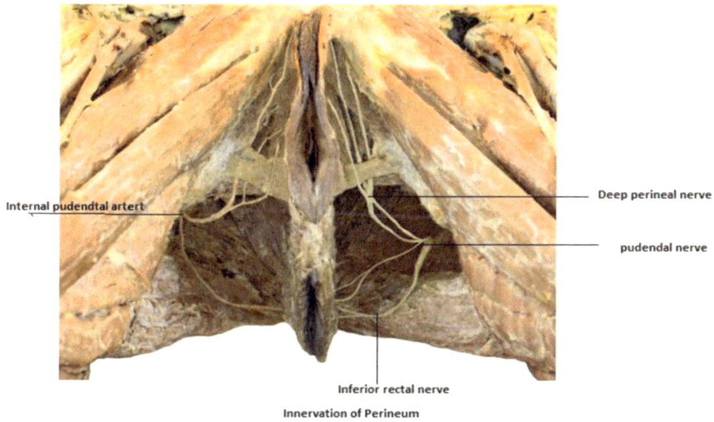

Figure 4.
Innervation of perineum.

respectively. The difference was significant after three vaginal deliveries, as compared to the first two (odds ratio, 3.2; confidence interval, 1.1–9.1) [25].

4.4 Blood and nerve supply of the region

The Pudendal nerve provides motor and sensory innervation of the perineum. Its origin is from S2 to S4 and exits the pelvis through the greater sciatic foramen, hooks around the ischial spine, and then travels along the medial surface of the obturator internus through the ischiorectal fossa in Alcock's canal. It emerges posterior and medial to the ischial tuberosity and divides into three branches: Clitoral, Perineal and inferior rectal nerve to supply the perineum [14]. The blood supply to the perineum is via the pudendal artery, which travels with the pudendal nerve to exit the pelvis. Pudendal artery also has three main branches with rich collateral anastomoses supplying the area (**Figure 4**).

5. Anal triangle

The anal triangle comprises of the anal canal, anal sphincter, perineal body, and ischiorectal fossa.

5.1 Perineal body

The perineal body is a fibromuscular structure that gives attachment to numerous structures and is located between the rectum and urogenital triangle. Anteriorly it is attached to the vaginal muscularis submucosa and the posterior border of the perineal membrane anteriorly where as pubovaginalis and puborectalis hold it in place anteriorly to the pubic bone. Laterally it is attached to the ischiopubic rami by superficial transverse perineal muscles. The perineal body communicates posteriorly and superiorly with the internal anal sphincter's longitudinal fibrous sheath, which begins to curve anteriorly towards the posterior wall of the vagina at the level of the anorectal junction [26].

Dissection and histological studies results have shown that the PB is the site of insertion of different pelvic floor muscles. In contrast to a previous description of a superior and inferior portion, Larson et al. with the help of MRI described three portions: an inferior portion, a midportion, and a superior portion. The inferior portion gives attachments to the bulbospongiosus muscle, the superficial transverse perineal muscle, and the external anal sphincter. The midportion comprises the insertions of the superior aspect of the transverse superficial muscles, the puboviscerаlis muscle (e.g., puboperinealis and puboanalis) as well as the internal and external anal sphincter muscles. The superior portion consists of the internal anal sphincter as well as some portion of the puboviscerаlis (e.g., pubovaginalis and puboanalis).

5.2 Rectovaginal facia (RVF)

The existence of the RVF has been a topic of debate for many years among anatomists and surgeons [27]. The main point of discussion is whether a distinct membranous fascial structure exists between the vagina and the rectum in women and whether the rectovaginal space is merely filled with a mix of adipose tissue, fragmented membranous connective tissue, and muscle fibers. The RVF has been found to be separate from the anterior rectal wall by a layer of adipose connective tissue that surrounds the rectum and allows the two organs to move independently. Evidence confirms that the RVF can be clearly visualized and measured with MRI [28] and the findings correlate with the dissection observations.

5.3 Anal canal

Surgically anal canal is approximately 4 cm long and the rectum terminates in the anal canal. It runs from anal verge to the anorectal ring referred as the proximal part of the levator–external anal sphincter (EAS) complex [29]. It lies between the ischiorectal fossae laterally, the anococcygeal ligament posteriorly, and the perineal body anteriorly. The anal canal can be divided into following three divisions:

1. The first section, 15 mm long, lined by a stratified cubic epithelium, extends from anorectal junction to the dentate line. The anal columns, which consist of 6–10 vertical folds, called the columns of Morgagni house a terminal radicle of the superior rectal artery and vein. Subepithelial tissues expand into three anal cushions. These cushions seal the anal canal and help maintain continence of flatus and liquid stools. The mucosal folds known as the anal valves, connect the lower ends of the columns, contain tiny depressions known as anal sinuses. Embryologically dentate line is the site of anal membrane which first appears at the confluence of the ectodermic (derived from proctoderm) and endodermic (derived from cloacae) and entodermic (derived from the cloacae) portions of the anal canal.

2. The transition zone, or pecten, is the second section and is 15 mm long. It is a lined by stratified, nonkeratinized squamous epithelium that lacks hair follicles, sweat and sebaceous glands, but is rich in somatic nerve endings. It corresponds to the intersphincteric anal groove and extends from the dentate line to the white line of Hilton.

3. The third portion, which is 10-mm long and continuous with the skin, is situated inferiorly to the Hilton white line.

The puborectalis muscle, the internal and external anal sphincters, and the sphincteric complex all surround the anal canal's walls. As a result of tonic circumferential contraction of the sphincter, the skin is arranged in radiating folds around the anus and is called the anal margin [30].

5.4 Anal sphincter complex

The anal sphincter complex is made up of the internal anal sphincter (IAS) and external anal sphincter (EAS) separated by the conjoint longitudinal coat. Both are different in structure and function even though they function as a single entity. When the rectum inserts into the pelvic diaphragm, the inner circular muscle of the rectum continues to the IAS and terminates approximately 1 cm proximal to the distal edge of the EAS. The IAS is found to be approximately 2–3 mm in thickness and to become thicker near the anal verge. Innervation for IAS comes from the inferior pelvic plexus which constitutes both sympathetic and parasympathetic fibers, splanchnic nerves (S2–4) also innervate this muscle, having a parasympathetic portion only. Due to this nerve supply, there is a constant resting pressure which may reflect the tone of the IAS [31]. With solid stool in the rectum, the IAS begins to relax, and the external anal sphincter begins to contract; this is called the rectoanal inhibitory reflex.

The EAS is a striated muscle that forms a muscular cylinder around the anal canal including the entire length of IAS and Conjoined longitudinal Muscle (CLM). It is innervated by the inferior rectal branch of the pudendal nerve. The EAS is considered to have one or more subdivisions, although the precise composition is disputed [32]. It consists of subcutaneous, superficial, and deep components. From a surgical viewpoint, these distinctions are less relevant. The deep portion of the external sphincter does not have posterior attachments. Its fibers are intimately related to the puborectalis muscle and insert anteriorly into the perineal body. The anococcygeal ligament is a dense connective tissue structure that attaches the superficial portion of the EAS to the tip of the coccyx. The subcutaneous part is circular but may have attachments to the perineal body anteriorly and the anococcygeal ligament posteriorly. In an endorectal ultrasound, the EAS appears as a hyperechogenic structure. In contrast with other skeletal muscles, the EAS maintains an unconscious resting electrical tone through a reflex arc at the cauda equina level, and thus generates 25–30% of the resting anal sphincter tone. Reflexive or voluntary contraction of the EAS/puborectalis works together to prevent fecal leakage. An active contraction can only be sustained for 30–60 seconds [33]. With solid stool in the rectum, the IAS begins to relax, and the external anal sphincter begins to contract; this is called the rectoanal inhibitory reflex.

5.5 The longitudinal layer and the conjoint longitudinal coat

The longitudinal muscle of the rectum fuses with striated fibers of the levator ani and puborectalis muscles at the level of the anorectal ring to form the conjoined longitudinal muscle. The fibers descend between the internal and external anal sphincters and consist of a fibromuscular layer, Distally, this muscle lamina becomes fibroelastic and divides into about 10 fibroelastic septa irradiating and crossing the subcutaneous part of the external sphincter to insert into the perianal skin forms the corrugator cutis ani [34, 35].

5.6 Innervation of the anal sphincter complex

IAS is a continuation of the circular fibers of the rectum, it shares the same innervation: sympathetic (L5) and parasympathetic nerves (S2–S4). The conjoint longitudinal coat is innervated by autonomic fibers from the same origin. The EAS is innervated by the inferior rectal branch of the pudendal nerve [32].

5.7 Vascular supply

The anorectum receives its major blood supply from the superior (terminal branch of the inferior mesenteric artery) and inferior haemorrhoidal (branch of the pudendal artery) arteries, and to a lesser degree, from the middle haemorrhoidal artery (branch of the internal iliac), forming a wide intramural network of collaterals. The venous drainage of the upper anal canal mucosa, IAS and conjoint longitudinal coat passes via the terminal branches of the superior rectal vein into the inferior mesenteric vein. The lower anal canal and the EAS drain via the inferior rectal branch of the pudendal vein into the internal iliac vein [33].

5.8 Lymphatic drainage

The anorectum is rich in lymphatic plexuses. The dentate line represents the interface between the two different systems of lymphatic drainage. Above the dentate line (the upper anal canal), the IAS and the conjoint longitudinal coat drain into the inferior mesenteric and internal iliac nodes [36]. Lymphatic drainage below the dentate line, which consists of the lower anal canal epithelium and the EAS, proceeds to the external inguinal lymph node.

5.9 Ischioanal fossa

The ischioanal fossa is a space filled with fat lateral to the anal canal and just below the pelvic diaphragm. It has a shape of a triangular pyramid with the apex at the boundary of the anal canal and the obturator fascia, and the base directed towards the perineal surface and is bound anteriorly by the perineal membrane, superiorly by the fascia of the levator ani muscle and medially by the EAS complex at the level of the anal canal. The lateral border is formed by the obturator fascia and inferiorly by a thin transverse fascia, which separates it from the perianal space. The ischioanal fossa contains fat and neurovascular structures, including the pudendal nerve and the internal pudendal vessels, which enter through Alcocks's canal [35].

6. Clinical considerations

6.1 Support to anorectum in pelvic organ prolapse (POP)

The sphincter vaginae (pubovaginalis) muscle passes through the perineal body and blends with the deep part of the external anal sphincter, anchoring the perineal body to the anal canal. In addition, the upper surface of the perineal body joins with the rectovaginal septum (Denonvilliers' fascia), and into its very posterior and superior aspects, pass fibers from the longitudinal muscle layer of the muscularis externa of the anterior wall of the rectum.

The perineal body is also attached to the muscles in the superficial and deep perineal spaces. Various components of the external anal sphincter, superficial transverse perineal muscles, levator ani (pubococcygeus, pubovaginalis, and puborectalis), and the median part of the posterior border of the perineal membrane are also attached to the perineal body making it to anchor the anterior anorectal region to the pelvic ring [24, 37–39].

An increase in the size of the urogenital hiatus is directly related to progressively larger pelvic organ prolapses that correlate more to the anterior-posterior (A-P) diameter than to the transverse diameter. Woodman [24] observed that it was actually the levator ani and not the perineal body that closes the urogenital hiatus. DeLancey and Hurd's (1998) data were consistent with this finding. The urogenital hiatus is sealed by the vaginal walls, endopelvic fascia, and urethra. As the urogenital hiatus increases in size, however, an open space develops in the pelvic floor and gravity tends to fill this space with a cystocele, rectocele, or the uterine cervix. Once the urogenital hiatus has opened up, the vaginal wall and cervix lie unsupported, and the endopelvic fascia is called upon to hold the pelvic organs in place. In predisposed individuals the constant load from abdominal pressure on this fascia can cause its failure with the development of significant prolapse. Ultimately, it is the perineal body that acts as the final mechanism for preventing prolapse beyond the urogenital hiatus. As the perineal membrane and tendinous arches of the pelvic fasciae tighten, the entire perineal body redirects abdominal pressure and dissipates gravity to prevent expansion of the urogenital hiatus.

6.2 Support to vagina

According to Woodman [24], the perineal body acts as Level III support to the immobile inferior one-third of the vagina. It serves to close the vagina and prevents its eversion.

He further described the perineal body s as the central connection between the two halves of the perineal membrane. As force is directed upon the vagina from above, the fibers of the perineal membrane, which are laterally and posteriorly attached to the ischiopubic rami and perineal body, respectively, become taut and resist downward displacement.

There is also structural interdependence between the support offered by the perineal body at Level III through its attachment to the perineal membrane and the suspension of the vagina by the perineal body at Level II (arcus tendinous fasciae pelvis and the levator ani in the middle one-third of the vagina) [4].

6.3 Obstetric injury to perineum

Perineal repair after childbirth affects millions of women across the globe. The prevalence of perineal trauma varies as it is dependent on obstetric practice including rates and types of episiotomies. Obstetric anal sphincter injuries (OASIS) occur in 1.7% (2.9% in primiparas) of women in centres where mediolateral episiotomies are practiced compared to 12–19% (19% in primipara) in centres practicing midline episiotomy [40].

It has been shown that up to half of OASIS are not recognized. Inadequate training of doctors and midwives in perineal and anal sphincter anatomy is believed to be a major contributing factor. However, despite recognition and primary repair of acute OASIS, 39–61% have symptoms of anal incontinence and 92% have persistent anal sphincter defects on ultrasound within 3 months of delivery [26]. The morbidity associated with perineal trauma depends on the extent of perineal damage, technique and materials used for suturing and the skill of the person performing the procedure.

6.4 Fecal continence

Because of its close relationship to the anal canal and anal sphincter muscles, the most important function of perineal body is to maintain fecal continence. The insertion and passage of internal and external sphincters into the perineal body is explained by Shafik [33]. Therefore damage to the perineal body causes direct (or indirect) extension into the anal sphincters and a loss of fecal continence is possible.

6.5 Urinary continence

The urinary continence mechanism is complex and involves the coordination of the involuntary detrusor and intrinsic urethral sphincteric muscles. The levator ani muscle and external urethral sphincter muscle which are under voluntary control function as backup mechanisms, The levator plate is also a dynamic backdrop for the normal continence mechanism as it resists posterior rotational descent of the vagina and overlying urethrovesical junction during levator contraction. As a result of childbirth and subsequent damage to pelvic musculature, posterior rotational descent of the bladder neck can also occur at times of increased intra-abdominal pressure. The urethrovaginal sphincter attaches to the perineal body and contributes to the external urethral sphincter. Neurologic and direct damage to the pelvic floor can occur during delivery, and subsequent loss of urethrovaginal sphincter function may be just enough to make a woman incontinent [12].

Benson summarized the neurogenic hypothesis of urinary incontinence. Partial denervation of Levator Ani can occur after childbirth through direct compression, stretch, and devascularization of the pudendal nerve resulting in levator ani atrophy and loss of urethral and bladder support which contributes to urinary incontinence [41]. Various studies described the factors associated with urinary incontinence and found out that spontaneous and deliberate trauma to the perineal body is associated with urinary incontinence and not episiotomy, but only perineal suturing continued to be associated with urinary incontinence [42, 43].

7. Conclusion

Female perineum plays an essential role in Pelvic organ prolapse, urinary and fecal incontinence and vaginal delivery. Therefore, knowledge of the regional anatomy and function is essential for management of disorders.

Conflict of interest

No conflict of interest.

Note of thanks

We would like to thank Dr. Peter Petros for his permission to use figures from his articles and book chapters [22].

Author details

Novera G. Chughtai*, Urooj Kashif, Samia Aijaz and Sumera Malik
Aga Khan University Hospital, Karachi, Pakistan

*Address all correspondence to: novera.chughtai@aku.edu

References

[1] Mahadevan V, Chandak P. The surgical anatomy of the perineum. BMJ Military Health. 2013;**159**(suppl. 1):i10-i14

[2] Fahmy MW, Sanganeria T, Anatomy MS. Abdomen and pelvis, superficial perineal space. In: StatPearls. Treasure Island (FL): StatPearls Publishing; Jan 2022

[3] Mirilas P, Skandalakis JE. Urogenital diaphragm: An erroneous concept casting its shadow over the sphincter urethrae and deep perineal space. Journal of the American College of Surgeons. 2004;**198**(2):279-290

[4] DeLancey JO. Structural anatomy of the posterior pelvic compartment as it relates to rectocele. American Journal of Obstetrics and Gynecology. 1999;**180**(4):815-823

[5] Hall MI, Rodriguez-Sosa JR, Plochocki JH. Reorganization of mammalian body wall patterning with cloacal septation. Scientific Reports. 2017;**7**(1):1-7

[6] Willaert W, PCeelen W, Pattyn P, Van Nieuwenhove Y. New insights into the surgical anatomy and embryology of the rectum: A review. Gastrointestinal Surgery Series: Colorectal Surgery. 30 Sep 2014:1

[7] Kromer P. Further study of the urorectal septum in staged human embryos. Folia Morphologica. 1999;**58**(1):53-63

[8] Mundy AR. True pelvis, pelvic floor and perineum. 39th ed. Gray's Anatomy. 2005:1357-1370

[9] Stein TA, DeLancey JO. Structure of the perineal membrane in females: Gross and microscopic anatomy. Obstetrics and Gynecology. 2008;**111**(3):686

[10] Roch M, Gaudreault N, Cyr MP, Venne G, Bureau NJ, Morin M. The female pelvic floor fascia anatomy: A systematic search and review. Life. 2021;**11**(9):900

[11] Martin BF. The formation of abdomino-perineal sacs by the fasciae of Scarpa and Colles, and their clinical significance. Journal of Anatomy. 1984;**138**(Pt 4):603

[12] Oelrich TM. The striated urogenital sphincter muscle in the female. The Anatomical Record. 1983;**205**(2):223-232

[13] Kotarinos RK. Musculoskeletal pelvic anatomy. In: Biomechanics of the Female Pelvic Floor. Cambridge, Massachusetts: Academic Press; 2016. pp. 53-87

[14] Thakar R, Fenner DE. Anatomy of the perineum and the anal sphincter. In: Perineal and Anal Sphincter Trauma. London: Springer; 2009. pp. 1-12

[15] Rahn DD, Bleich AT, Wai CY, Roshanravan SM, Wieslander CK, Schaffer JI, et al. Anatomic relationships of the distal third of the pelvic ureter, trigone, and urethra in unembalmed female cadavers. American Journal of Obstetrics and Gynecology. 2007;**197**(6):668-6e1

[16] Gosling JA, Dixon JS, Critchley HO, Thompson SA. A comparative study of the human external sphincter and periurethral levator ani muscles. British Journal of Urology. 1981;**53**(1):35-41

[17] Stecco C, Macchi V, Porzionato A, Tiengo C, Parenti A, Gardi M, et al. Histotopographic study of the rectovaginal septum. Italian journal

of anatomy and embryology. Archivio italiano di anatomia ed embriologia. 2005;**110**(4):247-254

[18] Li JR, Lei L, Luo N, Chen N, Xu HT, Hu X, et al. Architecture of female urethral supporting structures based on undeformed high-resolution sectional anatomical images. Anatomical Science International. 2021;**96**(1):30-41

[19] Delancey JO. Correlative study of paraurethral anatomy. Obstetrics and Gynecology. 1986;**68**(1):91-97

[20] DeLancey JO. Structural aspects of the extrinsic continence mechanism. Obstetrics and Gynecology. 1988;**72** (3 Pt 1):296-301

[21] DeLancey JO. Pubovesical ligament: A separate structure from the urethral supports ("pubo-urethral ligaments"). Neurourology and Urodynamics. 1989;**8**(1):53-61

[22] Petros P. The anatomy and dynamics of pelvic floor function and dysfunction. 2nd ed. The Female Pelvic Floor: Function, Dysfunction and Management According to the Integral Theory. 2007:14-50

[23] Richardson AC, Lyon JB, Williams NL. A new look at pelvic relaxation. American Journal of Obstetrics and Gynecology. 1976;**126**(5):568-571

[24] Woodman PJ, Graney DO. Anatomy and physiology of the female perineal body with relevance to obstetrical injury and repair. Clinical Anatomy: The Official Journal of the American Association of Clinical Anatomists and the British Association of Clinical Anatomists. 2002;**15**(5):321-334

[25] Ryhammer AM, Bek KM, Laurberg S. Multiple vaginal deliveries increase the risk of permanent incontinence of flatus and urine in normal premenopausal women. Diseases of the Colon & Rectum. 1995;**38**(11):1206-1209

[26] Kapoor DS, Thakar R, Sultan AH. Obstetric anal sphincter injuries: Review of anatomical factors and modifiable second stage interventions. International Urogynecology Journal. 2015;**26**(12):1725-1734

[27] Macchi V, Porzionato A, Stecco C, Vigato E, Parenti A, Caro RD. Histo-topographic study of the longitudinal anal muscle. Clinical Anatomy: The Official Journal of the American Association of Clinical Anatomists and the British Association of Clinical Anatomists. 2008;**21**(5):447-452

[28] Rodríguez-Abarca MA, Hernández-Grimaldo EG, la Fuente-Villarreal D, Jacobo-Baca G, Quiroga-Garza A, Pinales-Razo R, et al. Gynecological influencing factors on the rectovaginal septum's morphology. International Urogynecology Journal. 2021;**32**(6):1427-1432

[29] Catania VD, Randi B, Di Salvo N, Lima M. Functional anatomy of the pelvic floor and the anorectum. In: Ano-Rectal Endosonography and Manometry in Paediatrics. Champions: Springer; 2022. pp. 11-23

[30] Fritsch H, Brenner E, Lienemann A, Ludwikowski B. Anal sphincter complex. Diseases of the Colon & Rectum. 2002;**45**(2):188-194

[31] Lee JM, Kim NK. Essential anatomy of the anorectum for colorectal surgeons focused on the gross anatomy and histologic findings. Annals of Coloproctology. 2018;**34**(2):59

[32] Rociu E, Stoker J, Eijkemans MJ, Laméris JS. Normal anal sphincter

anatomy and age-and sex-related variations at high-spatial-resolution endoanal MR imaging. Radiology. 2000;**217**(2):395-401

[33] Shafik A. A new concept of the anatomy of the anal sphincter mechanism and the physiology of defecation. III. The longitudinal anal muscle: anatomy and role in anal sphincter mechanism. Investigative Urology. 1976;**13**(4):271-277

[34] Lunniss PJ, Phillips RK. Anatomy and function of the anal longitudinal muscle. British Journal of Surgery. 1992;**79**(9):882-884

[35] Morgan CN. The Surgical Anatomy of the Ischiorectal Space: President's Address

[36] Gami B, Kubba F, Ziprin P. Human papilloma virus and squamous cell carcinoma of the anus. Clinical Medicine Insights: Oncology. 2014;**8**:CMO-S13241

[37] Bump RC, Mattiasson A, Bø K, Brubaker LP, DeLancey JO, Klarskov P, et al. The standardization of terminology of female pelvic organ prolapse and pelvic floor dysfunction. American Journal of Obstetrics and Gynecology. 1996;**175**(1):10-17

[38] Siccardi MA, Bordoni B. Anatomy, abdomen and pelvis, perineal body. In: StatPearls. Treasure Island (FL): StatPearls Publishing; 2021

[39] Sánchez-Ferrer ML, Prieto-Sánchez MT, Moya-Jiménez C, Mendiola J, García-Hernández CM, Carmona-Barnosi A, et al. Anogenital distance and perineal measurements of the pelvic organ prolapse (POP) quantification system. JoVE (Journal of Visualized Experiments). 2018;**139**:e57912

[40] Minini G. Perineal and anal sphincter obstetric injury. In: Childbirth-Related Pelvic Floor Dysfunction. Champions: Springer; 2016. pp. 25-33

[41] Heit M, Benson JT, Russell B, Brubaker L. Levator ani muscle in women with genitourinary prolapse: Indirect assessment by muscle histopathology. Neurourology and Urodynamics: Official Journal of the International Continence Society. 1996;**15**(1):17-29

[42] Foldspang A, Mommsen S, Djurhuus JC. Prevalent urinary incontinence as a correlate of pregnancy, vaginal childbirth, and obstetric techniques. American Journal of Public Health. 1999;**89**(2):209-212

[43] Sleep J, Grant A. West Berkshire perineal management trial: Three year follow up. British Medical Journal (Clinical Research Ed.). 1987;**295**(6601):749-751

Section 3

Pearls Gallery of Clinical Practice

Chapter 3

Pelvic Floor Dysfunction

Yueying Chen and Ran Pang

Abstract

Pelvic floor disorders (PFDs) refer to a group of conditions comprising pelvic organ prolapse (POP), urinary incontinence, and urination or defecation abnormalities caused by weakened or damaged pelvic floor muscles, nerves, and connective tissues. The pelvic muscles, fascia, or ligaments around the pelvic organs, present an anatomical hammock across the pelvic floor. They play a vital role in supporting the normal function of pelvic organs. Once these muscles or connective tissues are damaged due to some reasons, the PFDs will consequently occur. Up to now, the diagnosis and treatment of PFDs are still filled with challenges due to the various etiology and symptoms of PFDs. In this chapter, we focus on urinary incontinence and POP, the two common conditions of PFDs, besides general introduction to PFDs. Not only is the latest evidence reviewed but also our viewpoint is presented. In general, PFDs involve urology, gynecology, anorectal department, and other fields, and the disciplines are usually associated with each other. Therefore, a multidisciplinary approach is usually necessary. With a developed understanding of the anatomy and function of the pelvic floor, pelvic floor reconstruction has received much attention and more new technologies are also constantly emerging, developed to improve the traditional methods.

Keywords: pelvic floor dysfunction, pelvic floor prolapse, urinary incontinence, urology, gynecology, proctology

1. Introduction

Pelvic floor dysfunctions (PFDs) are a common global health problem, which refers to highly prevalent conditions in women in which the pelvic floor muscles, nerves, fascia, or ligaments around the bladder, rectum, and vagina work abnormally. Pelvic floor dysfunction comprises pelvic organ prolapse (POP), urinary incontinence, urination or defecation abnormalities (e.g., overactive bladder or fecal incontinence), chronic pelvic pain, urogenital fistula, and female sexual disorders, which are widely concerned around the world. It involves urology, gynecology, anorectal department, and other fields, and the disciplines are usually associated with each other. The overall incidence of PFDs is difficult to evaluate precisely due to the wide range of disciplines involved. PFDs could affect up to 50% of women of childbearing years [1]. Approximately a quarter of American adult women have been reported suffering from at least one of these diseases, while 16% of men were diagnosed with

PFDs [2, 3]. Many studies showed that about 11% of women had received one or more surgical treatments for urinary incontinence or POP by the age of 80 years [4–6]. The risk for women who undergo surgical intervention related to stress urinary incontinence in their lifetime was 20.5% [7]. Pelvic floor dysfunction has a profound and long-term impact on patients, which not only increases the personal psychological burden and has a significant negative impact on the quality of life, but also seriously interferes with individual daily activities in social and professional fields. The comprehensive diagnosis and treatment of pelvic floor disorders are of rising interest and need due to an increasing age and the obesity epidemic, and the number of women affected by these conditions may continue to rise.

Breakthroughs have been made in anatomy and pathophysiology theory on pelvic floor dysfunction since 1990. Petros and Ulmsten proposed an Integral Theory (IT), which states that POP and abnormal pelvic symptoms (e.g., urinary stress and urge symptoms) mainly arise from lax connective tissues in the vagina or its supporting suspensory ligaments of the anatomic structures on the pelvic floor. Three organ groups, including bladder and urethra, vagina and uterus, rectum, and anus, are linked together by surrounding muscles and supporting ligaments. The normal opening and closure mechanism of these organs are able to be guaranteed by four important muscle groups (pubococcygeal muscle, levator plate, longitudinal muscle of the anus, and the puborectalis muscle) only when the suspensory ligaments are intact [8]. Functionally, these pelvic floor tissues were under the control of the same central nervous system, so the functional and dysfunction performance of any group of these organs are not isolated but correlated. As a result, muscles, ligaments, and nerves are used as links to integrate multiple disciplines such as urology, obstetrics and gynecology, and anorectal surgery into pelvic floor surgery. Functional urology is a subspecialty of urology that can solve a variety of common diseases, including urinary incontinence, POPs, sensory and voiding changes, defecation dysfunction, female sexual dysfunction, chronic pelvic pain syndrome, and fecal incontinence. Female urology, with a focus on the diagnosis and treatment of female pelvic floor dysfunction, has gradually sprung up and become a new subspecialty of urology. Much experience has also long been accumulated by obstetrician-gynecologists for the management practice of PFDs. However, PFDs have traditionally been conducted in a separate treatment by each relevant discipline, resulting in delays in health care, repeated counseling, the resultant significant public health burden, economic costs, and unsatisfactory efficacy [9]. Therefore, complete independence between disciplines is not conducive to the overall recovery for patients. It is gratifying to note that in recent years, urologists, anorectal physicians, and gynecologists have become increasingly aware of the need for closer communication and cooperation in treating PFDs, providing more support to each other to solve this troublesome problem. With the emergence of new technologies for pelvic floor reconstruction and rehabilitation, the mutual integration and cooperation of urology, gynecology, anorectal, pain, rehabilitation, and other departments in the treatment of PFDs are becoming more frequent. The causes and etiology of conditions related to PFDs are manifold but not fully unclear. The Integral Theory shows that the pelvic floor exists as an organic whole with interrelated anatomy and function rather than a simple superposition of every component, which is composed of supporting axes of different compartments, maintaining its internal stability and balance. When this balance is broken by any damage, other structures in the system can compensate for the function of the damaged structure to keep the system stable, but if the persistent injury exceeds a certain compensatory range, it will cause PFDs.

1.1 Risk factors of pelvic floor dysfunction

There are several risk factors that may be associated with an increasing prevalence of pelvic floor disorders over time. The risks higher with any of the characteristics are shown in **Box 1**.

Modifiable risk factors

- An increasing body mass index (BMI) over 25 kg/m^2
- Smoking habits
- Lack of exercise
- Constipation (a condition that causes an increase in abdominal pressure)
- Low estrogen status
- Diabetes

Non-modifiable risk factors

- Age (increased risk with aging)
- Race
- Related to gene
- Family history of urinary incontinence, overactive bladder, or fecal incontinence.
- Gynecological cancer and any treatment for PFDs before
- Gynecological surgery (e.g., a hysterectomy)
- Fibromyalgia
- Chronic respiratory disease and cough that lead to an increased abdominal pressure (e.g., chronic cough may result in an increasing risk of fecal incontinence and flatus incontinence).
- Innate dysplasia and acquired damage of the spinal cord.
- Menopause in women.
- Prostatectomy in men.
- Related to drug (such as inappropriate use of muscle relaxation drugs).
- Sexual abuse

Related to pregnancy:

- Maternal age (being over 30 years)

- Having delivered one or more babies before their current pregnancy.
- Having diabetes in the period of pregnancy.
- Multiparity

Related to childbirth: (pelvic floor muscle and nerve were damaged)

- Previous vaginal deliveries (forceps or vacuum).
- A vaginal birth when the baby is lying face up (occipital posterior).
- An active second stage of childbirth takes more than 1 hour.
- Injury to the anal sphincter during birth.

Box 1.
Risk factors for pelvic floor dysfunction.

1.2 Anatomy and physiology of pelvic floor and control mechanism of urine

The pelvic floor structure, a hammock-like shape, is composed of tough muscles, ligaments, and fascia. The levels between muscle, ligament, and fascia are complex, and the bladder, uterus, and rectum are close to each other. They cooperate and interact with each other, affecting the overall function of the pelvic floor system. The pelvic bone encircles the lower part of the abdominal-pelvic cavity, and the soft tissues seal the gap between the pelvic walls, thus forming the primary bottom tissue of the pelvic cavity. Bladder, urethra, uterus, rectum, and anus are involved in the pelvic floor. The support of the pelvic floor is vital for the urinary control mechanism of the lower urinary tract. The pelvic floor is a three-dimensional structure, which is viewed as a functional unit by more and more scholars in recent years, and the purpose of surgery should not be limited to the removal of lesions or the implantation of sling and other supports. In the field of urology, the target is to restore the normal anatomical relationships as much as possible, reconstructing the function of micturition and defecation. The characteristics of the anatomy and physiology of the pelvic floor differ between women and men (see **Figures 1** and **2**). In the following section, the mechanisms of urine control will be penetrated from different sex aspects, starting with anatomical and physiologic considerations.

1.2.1 Control mechanism of urine in female

Urinary control in women is a consequence of the combination of various factors, and the bladder, urethra, related nervous system, the surrounding suspensory ligaments, and supporting tissues play an essential role in urine control. The urethra is a flexible sealed pipe and mainly covered involved three anatomical tissues: the urethral mucosa, internal urethral sphincter, and external urethral sphincter, while the surrounding structures that support the bladder and urethra are the anal levator muscle

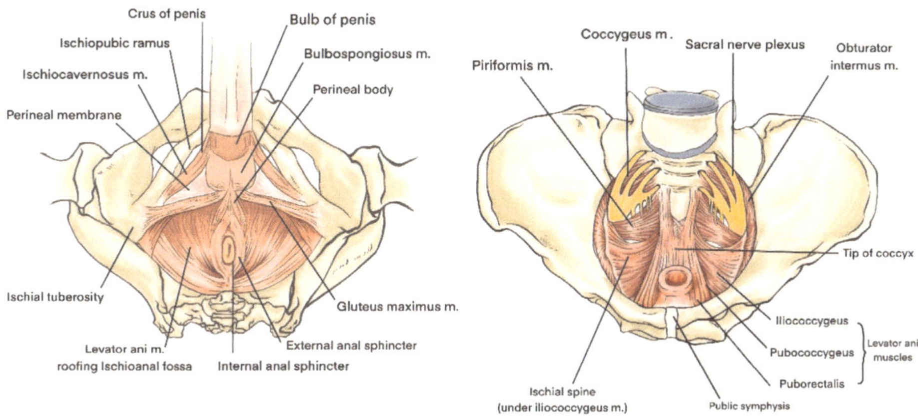

Figure 1.
Muscles of the male pelvic floor and digital palpation of deep pelvic floor muscles. M presents muscle. Adapted from Faubion et al. [10] and Rehder et al. [11].

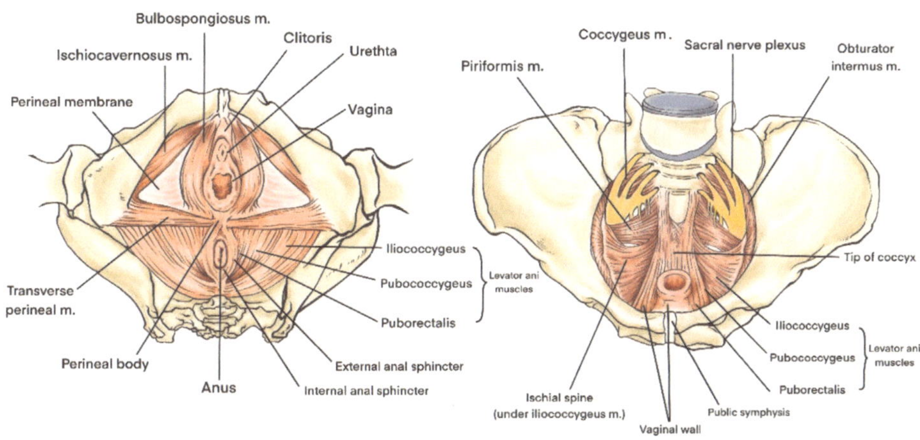

Figure 2.
Muscles of the female pelvic floor and digital palpation of deep pelvic floor muscles. M presents muscle. Adapted from Faubion et al. [10].

and the intrapelvic fascia. Urine control in women is mainly caused by the tension forced along the proximal urethra and/or the middle urethra. In addition, the distal urethral muscle also plays a certain role. Female urethral mucosa consists of loose connective tissue, a transitional epithelial tissue, which is associated with the urethral opening and closure. The urethral mucosa is abundant and often performs a formation of urethral mucosal ruffles. It allows for its expansion (opening during micturition) and helps to form the mucosal hermetic seal, which leads to an enhanced urethral closure reflex during bladder filling. The mucosa-submucosa of the urethra represents a spongy-shaped tissue and is abundant in elastic fibers and venous plexus. This property promotes a tension of the urethral mucosa and expansion of the urethral cavity, contributing to forming an effective water-tight seal. In addition, the urethral mucosa is sensitive to estrogen, which can increase the blood flow in the urethra and the vasodilation of blood vessels of the urethral submucosa. It may be one of the reasons why estrogen is able to improve the symptoms of patients with stress urinary incontinence. The periphery of the urethral submucosa is surrounded by the periurethral fascia. The effective function of the bladder neck originates from specific urethral mucosal properties, including internal sphincter (inherent characteristics of bladder neck/proximal urethra and distal (striated muscle) sphincter. The internal urethral sphincter is responsible for maintaining the closure of the bladder neck while the external urethral sphincter keeps the urethral closure pressure closed at rest. In cases of increased abdominal pressure, such as coughing and sneezing, the external urethral sphincter will contract to prevent the occurrence of urinary incontinence. At the same time, the levator anal muscle also contracts to support the bladder neck under the abdominal pressure. When the anal levator muscle is relaxed, the position of the bladder neck relative to the pubic bone is subsequently changed to voiding smoothly. In terms of the pelvic fascia, it can assist the levator anus in supporting and limiting the downward movement of the bladder neck.

1.2.2 Supporting structures of the female urethra

The supporting tissue of the urethra encompasses random muscle and connective tissue components. The full-length urethra is separated from the vagina by only one layer of fascia, and it is closely related to the anterior vaginal wall with a similar origin that both differentiates and develops from the urogenital canal. Strong muscle support (anterior vaginal wall) provides posterior support and middle urethral pressure. Relaxation of vaginal support can lead to anterior vaginal wall prolapse and adversely affect the urine control area. The musculature and fascia of the pelvic floor maintain a certain tension, further supporting the urinary control mechanism of women. The relaxation of pelvic floor caused by various factors may change the position and function of the bladder neck and urethra.

1.2.3 Structure and function of the female urethral sphincter

The smooth muscle sphincter and striated muscle sphincter are the major urethral sphincters that play a major role in the female urinary control mechanism. The circular striated muscle, localized in the proximal urethra and (or) the middle urethra, is responsible for the sphincter control. The smooth muscle of the posterior urethra can be divided into two layers: the inner layer is the longitudinal muscle and the outer layer is the circular muscle.

They are derived from the internal and external longitudinal muscles of the bladder to the urethra, respectively. The smooth muscle sphincter is a physiological rather than an anatomical sphincter that is innervated by autonomic nerves and is under nonconscious control, keeping the bladder neck and proximal urethra closed during the entire storage period of urine. There is no complete and separate anatomical structure of the internal sphincter. The urinary control mechanism of the internal sphincter is not located at a certain site or level but involves the full length of the female urethra. External urethral sphincter muscle (EUS) refers to the muscle fibers of the urethral striated muscle sphincter being in circular permutation around the membranous urethra, which starts from deep transverse perineal muscle. Part of the muscle fibers fold into the proximal urethra, go up on the outside of the longitudinal muscle in the urethra, then stop at the middle part of the proximal urethra, and finally are connected to the outer circular muscle of the urethra. These fibers are called the striated sphincter muscle in the wall of the urethra, composed of slow-response muscle fibers. They can produce almost complete circular compression in the middle urethra under the influence of the pudendal nerves and maintain a certain tension for a long time without fatigue [12]. However, the sphincter muscle around the urethra contains not only many slow-response muscle fibers but also fast-twitch muscle fibers in which the fast-twitch muscle fibers are innervated to quickly contract and close the urethra, but they cannot maintain the closed state of the urethra for a long time. In most cases, the above additional forces, the inherent characteristics of the urethra and muscle elements that promote urine control led to urinary incontinence in women under a combined action of alone or a combination.

1.2.4 Anatomy of the pelvic floor and control mechanism of urine in male

The structures that contribute to male urinary continence include the bladder neck (preprostatic sphincter), prostate, striated urethral sphincter, and levator ani muscle. A comprehensive understanding of periprostatic anatomy is crucial to the successful performance of urological pelvic surgery.

1.2.5 Urethral sphincter

The urethral sphincter consists of internal sphincter and external sphincter (i.e., the smooth muscle urethral sphincter, the rhabdosphincter, preprostatic sphincter, and levator ani muscle). The internal sphincter encompasses smooth muscle fibers and elastic tissues, which may be irreparably damaged during the prostatectomy. There are unmyelinated nerve fibers filtering the smooth muscle aspect of the proximal urinary tract and the bladder neck at 5 and 7 o'clock. The external urethral sphincter is regarded anatomically as intrinsic striated urethral sphincter (rhabdosphincter), as well as striated slow-twitch fibers. It presents a close bond to the anatomy and function of the pelvic floor, which is responsible for urinary continence during prostatectomy. It encircles the sphincteric part of the urethra in a round shape or omega shape. The anterior inserts on the distal ventral layer of the prostate and the perineal fascia, form a dorsal midline raphe. Then, the midline raphe fuses with the posterior layer of the prostatic apex and Denonvilliers fascia and with the central perineal tendon. The muscle fibers are controlled by the inferior hypogastric plexus, which covers both sympathetic and parasympathetic fibers. The neurovascular bundle (NVB) is visible running with part of these fibers. There are myelinated nerve fibers penetrating the striated sphincter on either posterolateral side at 3 and 9 o'clock [13–15].

1.2.6 Levator ani and puboperinealis muscles

The levator ani is the innermost muscle of the pelvic floor, having abundant type II fast-twitch and slow-twitch fibers. Levator ani encircles the urogenital hiatus at its extreme near-striated urethral sphincter. At the hiatus, it links to the striated urethral sphincter and thickens. When an increased pressure in the intra-abdominal suddenly presents, contraction of this thickened levator sling at the urogenital hiatus may show a crucial contribution to urinary continence. The striated sphincter at the perineal body originates from puboperinealis and runs along the lateral aspect of the prostate and the sphincteric part of the urethra, then meets on the dorsal aspect of the perineal body. These mainly are fast-twitch muscle fibers that contribute to quitting urine flow by angulating the urethra. The anteromedial fibers of the levator ani muscle are involved in the continence mechanism due to their strong connection with the rhabdosphincter and the prostate. Prostate apical dissection and urethrovesical anastomosis may result in the damage of these structures with the resultant urinary incontinence [15].

1.2.7 Supporting structures of the male urethra

The anterior urethral attachments are composed of pubovesical and puboprostatic ligaments, as well as the tendinous arch of the pelvic fascia. It can stabilize the urethra behind the pubis. The posterior support of the sphincteric urethra depends on a complex combination of Denonvillier's fascia, the rectourethralis muscle, the perineal body, and the levator ani [13].

1.2.8 Bladder neck and prostatic urethra

The effective bladder neck is derived from specific urethral mucosal properties, containing the internal sphincter (the inherent characteristics of the bladder neck / proximal urethra) and the distal (striated muscle) sphincter, the male bladder neck has the internal sphincter passing through and extending to the distal seminal caruncle. The tension produced by the bladder neck is an important contribution to keep the body from incontinence. Without disease or injury, urine can usually be maintained above the bladder neck, not falling to the external sphincter level. On the one hand, the adrenergic in smooth muscle from the bladder neck and prostate urethra is activated. On the other hand, the sympathetic nerve distributed in this area as well as the prostate itself, further increases bladder neck resistance. Therefore, the complete loss of sphincter function may not lead to urinary continence in men.

1.3 History

A general history should be taken of current and past symptoms or disorders related to pelvic floor dysfunction, such as

- Urologic: urinary hesitancy/frequency/urgency, dysuria, bladder pain, urinary incontinence, and emptying disorders of the bladder

- Gynecologic: dyspareunia (during/after intercourse), sexual dysfunction, bulging from the vagina, and uterus/vaginal prolapse.

- Colorectal: constipation, fecal incontinence/leaking, rectal/anus prolapse, emptying disorders of the bowel, and bloating.

- Chronic pelvic pain/pressure: for female patients, abnormal work in pregnancy and childbirth is associated with potential symptoms of pelvic floor dysfunction. Ask women about the number of their previous pregnancies and giving birth, who have recently given birth (natural childbirth/cesarean derivation)about symptoms of PFDs during routine postnatal. For men, when asking them, pay attention to their potential prostate or urethral problems. Information on past history is also required to be detailed, such as the patient's previous history of pelvic floor surgery, gynecology and obstetrics, trauma and drug therapy, radiotherapy, and family history of POP, all of which may provide important information for clinical diagnosis and treatment [16]. A focused medical history should be interrogated around the symptoms described above, clinical examination and investigations to exclude other causes, such as pelvic masses, neurological diseases, urinary tract infections, diabetes, inflammatory bowel or bladder conditions, and fistula. Also, the history of symptoms should be carried out around the nature, severity, duration, and frequency of urinary incontinence, POPs, as well as other symptoms related to pelvic floor dysfunction. Besides, we should pay attention to the patient's quality of life affected by PFDs.

1.4 Physical examination

Physical examination of patients with PFDs encompasses general situation, abdominal and pelvic examination.

1.4.1 General condition

Age, body size, gait, and physical fitness are the factors that can reflect the mental state, neurological, and physical state of patients.

1.4.2 Abdominal exam

Combined with medical history, the abdominal exam can help to understand the patients' history of abdominal surgery and make an adequate evaluation of the incision, hernia, abnormally enlarged organs, and abdominal morphology.

1.4.3 Pelvic exam

At least seven key items are included for female pelvic examination: Digital rectal examination (DRE), external genitalia, urethral orifice, urethra, bladder, vagina, cervix, uterus and its appendages and para-uterine tissues, anus, and perineum.

- Digital rectal examination (DRE): to assess the anal sphincter tone, rule out malignancy, and locate the origins of pain, hemorrhoids, anal fissures, and anorectal abscesses. In men, the size, nodularity, and tenderness of the prostate can be evaluated.

- External genitalia, perineum: inspect general appearance, color of the external genitalia, labia for size and adhesion, and vulva and vagina for atrophy; to evaluate female estrogen level. For example, the vulva is pink and wrinkled under normal estrogen levels. In contrast, the vulva is pale, flat, dry, and unwrinkled at estrogen-deficient level.

- Urethra: cotton swab test for localizing vulvodynia and assessing the position and mobility of urethra; excessive mobility of the urethra refers to the angle of cotton swab changes exceeds 30°C [17, 18].

- Vagina, cervix: speculum exam for atrophy or inflammation of the vaginal mucosa and visualization of the cervix.

- Anal: sensation, response to touch and pinprick.

- Perineum: graves speculum for the degree of the perineal relaxation in the lithotomy and standing position.

- The pelvic organs and floor muscles: bimanual exam/digital muscle testing for the strength of pelvic function muscle (PFM) and pain; palpation of the urogenital triangle encompasses ischiocavernosus, bulbospongiosus, and transverse perineal muscles and perineal body, evaluating dyspareunia.

In men, the assessment of penile and urethral orifice for stenosis is included in the exam, especially for men with urinary incontinence, following prostatectomy. The examination of urinary incontinence is best performed in a standing position. In addition, a neurological exam is important for patients with known or suspected neurological diseases. Make an evaluation of the patients' mind, speech, cognitive function and gait, deep and superficial sensations in the lower extremities, perineum and perianal areas, nerve reflexes, and tension on vaginal and pelvic floor muscle, which is helpful for the diagnosis and treatment of PFDs. The bulbocavernosus reflex (BCR) and anal sphincter tension represent the neural function in S2–4 segments. A positive BCR means that when the glans or clitoris is squeezed, the anus and pelvic floor will contract.

1.5 Evaluation

In addition to the above-detailed history and physical examinations, there are various other supplementary measures used to evaluate the pelvic floor function, symptoms, and quality of life of patients, such as

- Questionnaire: Urinary diaries, UDI-6 and ICIQ-UI (widely used to assess the related pelvic floor symptoms of urinary incontinence and non-incontinence), pelvic floor dysfunction-related quality of life scale.

- Pad test/weights: An accurate evaluation of urinary incontinence and quantifies urinary loss.

- Urine analysis: The guidelines in the Society of Urodynamics, the American Urological Association (AUA), and Female Pelvic Medicine and Urogenital Reconstruction (SUFU) mentioned that urine analysis should be performed in patients with urinary incontinence to exclude infection, micro-hematuria, proteinuria, and glucosuria [19, 20].

- Urodynamics: To measure the functional aspects of the distal urinary tract, including urine storage and voiding.

- Cystoscopy: Visual inspection of the bladder and urethra to exclude suspected bladder or urethral pathology.

- Ultrasonography: To measure the postvoid residual volume (PVR); if necessary, endoanal ultrasonography can be performed to evaluate the detection of sphincter defects, infectious process, or neoplastic diseases of the anal canal, excluding a traumatic defect related to fecal incontinence; pelvic ultrasound to assess the uterus and adnexa; dynamic transperineal ultrasound is focused on exploring some functional diseases of the pelvic floor.

- CT scan MRI of the abdomen and pelvis: To evaluate structural anatomy.
- Electromyography (EMG): Electrodes (needle or surface) measure external sphincter activity during contraction and relaxation.
- Endoscopy: (Anoscopy, sigmoidoscopy, colonoscopy), incorporated as indicated to evaluate for other sources of pelvic pain (colorectal, gynecologic, neurologic, orthopedic, and urologic).
- X-ray/MR defecography: Defecography allows the evaluation of the defecation process in the physiological sitting position with the advantage of detailed morphological and dynamic evaluation of the pelvic floor structures and pelvic organs in a noninvasive way. It can provide an evaluation for rectal prolapse, rectocele, enterocele, and perineal descent.

Besides, uroflowmetry, excretory urography or cystourethrography, anorectal manometry, and balloon expulsion can be considered in specific cases. An integrated test approach is needed in the assessment of these patients, adequate to understand the complex anatomy and function of the pelvic floor.

1.6 Management

With initial assessment in primary care, therapeutic management for patients with PFDs should be planned to accommodate their specific needs. A multidisciplinary team approach for the management of pelvic floor dysfunction is often necessary. The common therapies are performed in **Box 2**.

Conservative management of pelvic floor dysfunction

- Lifestyle changes
- Weight loss: advise for patients with a body mass index (BMI) over 30 kg/m^2
- Diet: appropriate intake of alcohol and caffeine (cola, tea, and coffee) for overactive bladder or urinary incontinence; reduction or avoidance of cigarettes, acidic foods/beverages, including citrus and tomatoes, concentrated sugar, artificial sweeteners, and spicy foods [23].
- Physical activity: walking, swimming, and yoga.
- Pelvic floor muscle training (PFM): pelvic floor exercises (Kegel) for strengthening the pelvic floor. For patients who cannot finish an effective pelvic floor muscle contraction, consider biofeedback techniques, electrical stimulation, or vaginal cones as supplementing pelvic floor muscle training.
- Behavioral approaches: bladder retraining, toileting, and habit training.
- Psychological interventions.

- Medicines: Topical vaginal estrogen for overactive bladder, vaginal thinning, and dyspareunia; anticholinergics (fesoterodine, tolterodine), beta3 agonists (mirabegron) for overactive bladder.
- Manipulation: Pessaries for symptomatic POP and urinary incontinence, biofeedback, physical therapy (trigger point massage, myofascial release, strain-counterstain, joint mobilization)
- Invasive procedures: Botulinum toxin A injections to the pelvic floor, electric neuromodulation (sacral neuromodulation (SNM) /percutaneous tibial nerve stimulation (PTNS)), and trigger point injections or acupuncture for pain [24].

Surgical management of pelvic floor dysfunction

- A surgical approach is necessary if patients do not have satisfactory symptom relief with nonsurgical treatments.
- Urinary incontinence: mid-urethral sling (MUS) surgery (e.g., tension-free vaginal obturator route tape (TVT-O), transobturator tape (TOT), tension-free vaginal tape (TVT)); (open or laparoscopic) colposuspension; urethral injections; autologous fascial sling the pubovaginal sling (PVS); artificial urinary sphincter (AUS).
- Pelvic organ prolapse (POP):
    - Anterior vaginal wall prolapse (i.e., cystocele, urethrocele, and paravaginal defect): anterior vaginal wall repair, transvaginal paravaginal repair, colposuspension.
    - Posterior vaginal wall prolapse
    - Enterocele: Repair of the rectovaginal fascia and obliteration of the cul-de-sac.
    - Rectocele: Posterior vaginal wall repair or transanal repair, rectopexy (posterior or anterior), or perineal resection (Altemeier).

Box 2.
Management of pelvic floor dysfunction [21, 22].

Regarding invasive procedures, botulinum toxin A(BoNT-A) and urethral bulking agents (UBAs) are normally performed as outpatient procedures to save costs. Growing evidence has shown that BoNT-A is a promising alternative option for treating UUI/OAB/CPP and UBAs have benefits for SUI in certain specific cases, such as patients who are fragile, have severe comorbidities or high anesthetic risk, are unwilling to undergo more invasive procedure, or have limited surgical options (postoperatively, after irradiation) [23, 25].

Besides, the health care model of PFDs has drawn growing attention. There have been a number of different models available for the management of PFDs, but the most effective one has not yet been determined. Nurse-led primary healthcare has become a topic gradually gaining attention in female PFDs. Traditionally, primary health care nurses and general practitioners are typically tasked with diagnosing and managing elderly patients with urinary incontinence, OAB, prolapse, and other chronic illnesses. Nurse-led care services in primary health care for community-dwelling people are in high demand, especially for frail patients, post-stroke survivors, and dementia individuals. Nurses play a crucial role on the health care team in evaluating and treating PFDs. Not only is the patients' health status monitored but also advice about the treatment and referrals is offered, which may support the patients in achieving the agreed-upon care and analyzing the results of the care. The findings of a study indicated that a nurse-led care protocol (SICP) based on the star model can significantly decrease incontinence episodes and urine leakage and improve the quality of life in women with SUI [24]. Oliver and his colleagues assessed the effectiveness of a healthcare service model called a nurse-led urology triage clinic (UTC) and they found that the UTC could result in a decrease in clinic visits and patient journey time. Furthermore, 59% of patients can be adequately managed by a specialist nurse [26]. Additionally, a RCT has revealed that a nurse-led telephone-based intervention before the initial urogynecology appointment shortened the duration of the appointment and enhanced the satisfaction of both patients and physicians [27]. A Dutch study showed that the implementation of a new nurse-led service for urinary incontinence led to better health outcomes in elderly community dwellers (2592–2618 QALYs gained in 2030) and lower cost-savings (€32.4 -€72.5 million in health care perspective; €182.0–€250.6 million in societal perspective) [28].

Because urinary incontinence and POP are two of the most common PFDs, this chapter focuses on urinary incontinence and POP.

Apical pelvic organ prolapse (i.e., uterus, vaginal vault): laparoscopic/robotic sacrocolpopexy; hysterectomy and uterosacral ligament suspension (apical repair), sacrospinous ligament fixation (SSLF), colpocleisis, and vaginal fornix suspension.

2. Urinary incontinence

2.1 Introduction

Urinary incontinence is defined as any involuntary loss of urine by the International Continence Society (ICS). It is a common severe health problem that impacts the quality of life and even economic consequences for millions of people. At present, the symptoms related to urinary incontinence are named by the International Continence Society (ICS)/ International Urogynecological Association (IUGA), including stress urinary incontinence, urge urinary incontinence, mixed urinary incontinence, persistent urinary incontinence, positional urinary incontinence, undetected urinary incontinence, coital incontinence, urgency, and nocturnal enuresis. The occurrence of urinary incontinence is closely associated with advanced age, pregnancy, delivery of babies, high-impact physical activities, hormone levels, body mass index (BMI), immobility, urinary tract infections, diabetes mellitus (DM), depression, smoking, prostate-related surgery, and other identifying well-established factors. Urinary incontinence (UI) affects up to 32% of men and at least 1/4 of women in their lifetime, but it presents more often in women [29–32]. Major causes of urinary incontinence in men are usually associated with following radical prostatectomy (RP), prostate irradiation, and transurethral resection of the prostate (TURP), affecting up to 69% of men [33]. It may damage the nerves, blood supply, and/or structures around the pelvic floor, resulting in weakening urethral sphincter function and urethral bulb [11].

2.2 Management of urinary incontinence

The treatment of urinary incontinence should be tailored to the patients for different causes and physiological and pathological changes. At first, based on a detailed examination, the clinician must determine whether the cause of the symptoms is the bladder itself or the bladder outlet problem or both. Then provide counseling advice and develop personalized programs for patients according to patients' needs, goals and expectations, risk-benefit ratio, and cost-benefit ratio. It is very important to provide patients with treatment goals that meet their expectations and achievability. Treatment of urinary incontinence contains nonsurgical and surgical treatments, and it is important to differentiate between different pathophysiological forms.

2.2.1 Conservative therapy of urinary incontinence

2.2.1.1 Lifestyle changing

Examples of lifestyle factors related to urinary incontinence encompass obesity, smoking, physical activity, and inadequate nutrition. Changing these lifestyles (i.e., weight loss, quitting smoking, reducing caffeine/fluid intake, dietary/behavioral modification, and reducing unnecessary medication) may help to improve urinary continence.

2.2.1.2 Pads usage

Patients with cognitive impairment or temporary urinary incontinence after a comprehensive assessment can be offered urinal pads.

2.2.1.3 Constipation improvement

Some studies have shown that constipation is strongly close to urinary incontinence and overactive bladder. Improving constipation through behavior, diet, and medication can contribute to treat urinary incontinence.

2.2.1.4 Behavioral therapy, physiotherapy, and electrophysiological stimulation of the pelvic floor

Bladder training is the first-choice treatment for female patients with urge incontinence or mixed incontinence. The European Society of Urology (EAU) and American Association of Urology (AUA) guidelines recommend pelvic floor muscle therapy (PFMT) as first-line therapy for female patients with stress urinary incontinence and mixed urinary incontinence, as well as men after prostatectomy [34]. Pelvic floor strength should be monitored and pelvic floor training should be continued for at least three months [35]. Besides, if the symptom improvement is not obvious after receiving anticholinergic drugs, female patients with urge incontinence may be considered to accept transcutaneous tibial nerve electrical stimulation (TTNS). In addition to pelvic floor training, electrostimulation (skin, vaginal, anal) may be provided to treat urinary incontinence. However, female patients with stress urinary incontinence or overactive bladder syndrome are not recommended to use electrostimulation alone. A systematic review indicated that PFMT combined with EMG-BF achieves better outcomes than PFMT alone in treating stress urinary incontinence or pelvic floor dysfunction [36]. Moreover, vaginal pessary therapy or the insertion of urethral plugs is an alternative conservative option for female patients with stress urinary incontinence. Women who are pregnant should be provided with prepartum physiotherapy that may decrease the risk of developing urinary incontinence.

2.2.2 Drug therapy for urinary incontinence

The appropriate drugs should be selected according to the type of urinary incontinence.

2.2.2.1 Antimuscarinic agents

If conservative non-pharmacological interventions fail to improve the symptoms of adults with overactive bladder or urge incontinence, then antimuscarinic agents must be additionally used. Antimuscarinic agents of extended-release formulations are an effective oral therapy option for female patients with urge incontinence. The common antimuscarinic agent contains solifenacin, tolterodin, oxybutynin, among others. If treatment with an antimuscarinic agent is not satisfactory, an increasing dose or the combination of anticholinergic agents with different components or mirabegron may be considered. Besides, changing the medication within the same group of medications or the form of delivery may be an effective option. Antimuscarinic agents, except for oxybutynin, are safe in the treatment of older patients. Oral oxybutynin has an obvious negative impact on cognitive functioning. Increasing awareness is needed to review the (total) anticholinergic burden, potential interactions with other medications, and possible intolerances.

2.2.2.2 β3-adrenergic agonists

β3-adrenergic agonists (mirabegron) are suitable for overactive bladder, urge incontinence, or mixed urinary incontinence. In terms of female patients who did not have any effective improvement or cannot tolerate antimuscarinic agents or in whom antimuscarinic agents are contraindicated, β3-adrenergic agonists (mirabegron) should be provided to them. Mirabegron has an equivalent effect of administration compared with antimuscarinic agents regarding the improvement of the symptoms of overactive bladder but without an increased risk of postvoid residual urine. It is important to have an awareness that one negative side effect of β3-adrenergic agonists (mirabegron) can be the risk of increasing blood pressure. β3-adrenergic agonists (mirabegron) combined with antimuscarinic agents may receive more benefits than escalating the dose of the antimuscarinic agent in female patients with an unsatisfactory treatment.

2.2.2.3 Duloxetine

If conservative therapy has not led to the desired success in patients with stress urinary incontinence, urge urinary incontinence, or in whom surgery is inappropriate, then duloxetine can be offered to patients. However, it should be noted that duloxetine has a high side effect.

2.2.2.4 Estrogens

Vaginal estriol administration should be considered in postmenopausal women with mixed urinary incontinence and vulvovaginal atrophy [37, 38]. The treatment of estrogens must be continued over long time periods.

2.2.2.5 Desmopressin

Desmopressin is currently not approved for use in urinary incontinence. Although it can relieve the symptoms in patients with urinary incontinence within a short period, long-term use is not recommended. It is essential to be aware that patients should be followed for the risk of hyponatremia before and during desmopressin therapy.

2.2.3 Surgical intervention for urinary incontinence

The choice of surgical intervention should be individualized based on the patient's symptoms, the severity of the patient's distress, the impact on quality of life, the patient's goals and expectations, the surgeon's experience, the medical and surgical history of patients, the availability of materials, and the potential for adverse events.

2.2.3.1 Surgical treatment of stress urinary incontinence

Surgical therapy should only be offered if patients with stress urinary incontinence have been treated with exhausted conservative therapies.

2.2.3.2 Surgical treatment of uncomplicated stress urinary incontinence

The characteristics of uncomplicated stress urinary incontinence contain no history of incontinence or prolapse surgery, no neurological symptoms, no concomitant symptomatic genital prolapse, or no wish to continue fertility. Slings have gradually become the standard therapy for women with stress urinary incontinence due to their various materials, implantation methods, and anchoring techniques. Several different surgical procedures have been reported for SUI, including mid-urethral sling (MUS), (Burch) colposuspension, autologous fascial sling, and the pubovaginal sling (PVS). Nowadays, MUS is considered as the surgical gold standard of stress urinary incontinence [39]. The MUS can be conducted through retropubic or transobturator routes with three types: retropubic MUS, such as retropubic tension-free vagina tape (TVT); transobturator MUS, such as transobturator inside-out tape (TVT-O) and outside-in tape (TOT); single incision sling (SIS), creating a dynamic mid-urethral support with synthetic material tapes. Compared with retropubic MUS, the benefits of transobturator MUS include a lower occurrence of bladder or vaginal perforation in the operation, major subsequent vascular complications and pelvic hematoma, a lower rate of supra-pubic pain, urinary tract infection, dysfunctional voiding, and lower urinary tract symptoms (LUTS) [40]. However, another study has shown that transobturator MUS presents a higher rate of recurrent procedures and a higher risk of groin pain than retropubic MUS [41]. Compared with retropubic or transobturator MUS, SIS can shorten operative time, have less intense postoperative, intraoperative bleeding, and a lower risk of postoperative voiding dysfunction. The currently available SIS may be appropriately offered to elderly women with anxiety of operative pain or postoperative voiding dysfunction [42]. If female patients wish to avoid surgical morbidity, UBAs are feasible, but the benefits must be balanced against recurrent injections and long-term results remain inadequate.

2.2.3.3 Surgical treatment of complicated stress urinary incontinence

Complicated stress incontinence is characterized by a prior history of failed anti-incontinence surgery, pelvic irradiation, neurogenic bladder, and concomitant POP. Recurrent urinary incontinence may be considered for reoperation with an artificial material mid-urethral sling, colposuspension, or autologous fascial sling. Compared with the first surgery, reoperation may have a higher risk of complications and a lower improvement. In general, the artificial urinary sphincter (AUS) and adjustable continence therapy (ACT®) balloons implantation are used for complicated stress urinary incontinence, but they have a high risk of complications, device failure, or the need for removal. As an alternative treatment, UBAs are a minimally invasive approach for select females with recurrent or persistent SUI after failed MUS procedures [25]. As a well-established treatment option, the usage of UBAs is increasing over the past few years [43–45]. Recently, a number of UBAs have been released, which include Bulkamid®, Macroplastiue®, Durasphere®, Coaptite®, or Urolastic® [46]. In general, UBAs can be performed as an office-based procedure under local anesthesia, which presents an encouraging short-term and acceptable medium-term efficacy, excellent safety profile, and few adverse events [46, 47]. The common adverse events include injection site rupture, urethral prolapse, urethral erosion, particle migration, and periurethral abscess. Unfortunately, there have been no studies comparing UBAs with other procedures for female SUI. In contrast to mid-urethral sling, the efficacy of UBAs seems to be less but can be balanced by a lower complication rate [48].

2.2.3.4 Surgical treatment of urge urinary incontinence

Onabotulinum toxin A (Botox) injection may be a minimally invasive option when conservative and drug treatment has failed in patients. It was approved for the treatment of adult refractory overactive bladder in 2020 [23]. Several studies have demonstrated that Botox is typically given in doses of 100 to 300 IU every 6 to 12 months and has been effective for 19 to 24 weeks after each injection. In clinical practice, it should be focused on the efficacy of onabotulinum toxin A for a limited time, as well as the risk of urinary tract infections and residual urine, and the possible need for intermittent self-catheterization. Generally, it is performed as an office-based procedure under local anesthesia, which avoids unnecessary costs associated with hospitalization. Sacral neuromodulation (SNM) should be considered if antimuscarinic agents have not led to the desired success in female patients with urge urinary incontinence. The option to offer augmentation cystoplasty should be considered in selected cases if the above methods are ineffective. Besides, urinary diversion should only be considered in patients when who have failed to respond to minimally invasive treatment, accept an artificial stoma, and with a low tumor risk. Detrusor myectomy to treat urinary incontinence is not recommended to patients.

2.2.3.5 Surgical intervention for male urinary incontinence

The artificial urinary sphincter (AUS) is defined as the gold standard therapy for postprostatectomy stress urinary incontinence. If male patients do not benefit from prior male urethral sling placement and gradually present with persistent or worsening stress urinary incontinence, the artificial urinary sphincter is better offered than a secondary sling [49]. Patients with severe postoperative urinary incontinence may be served with fixed slings. BTX-A injection is an excellent option for patients with small prostates and low prostate-specific antigen (PSA) levels, presenting better effects and tolerability. Although BTX-A injection has a positive effect on overactive bladder in men, evidence-based guidelines are still scarce [50].

2.2.3.6 Surgical treatment of mixed urinary incontinence

In patients with MUI, it is essential to identify whether urge or stress incontinence is the predominant symptom. The most troubling symptoms should be treated first. It is demonstrated that the overall cure rate with surgery is higher in patients with predominant symptoms of stress urinary incontinence than those with urgency-predominant symptoms of mixed urinary incontinence [51, 52]. Compared with a mid-urethral sling alone, a combined treatment with a mid-urethral sling and pelvic floor muscle therapy may result in slightly better outcomes [53].

2.2.3.7 Additional surgery for urinary incontinence

In recent years, various types of mesenchymal stem cells (MSCs) have been used to treat stress urinary incontinence due to their paracrine function, which contributes to tissue repair. There is some data supporting the feasibility and the safety of MCS therapy [54]. Besides, nerve growth factor (NGF) can contribute to nerve and muscle regeneration, and the study has investigated that a combination of stem cells and nerve growth factor (NGF) is more effective in relieving stress urinary incontinence than stem cell therapy alone [55]. Combination therapy is becoming an emerging and promising method for the treatment of urinary incontinence with a restoration of sphincter function. However, this stem cell therapy has limitations for clinical use due to the unclear potential mechanisms of action of stem cell therapy for SUI. And the extraction process as well as *ex vivo* expansion is still difficult. In addition, laser therapy may now be an option recommended to some women with stress incontinence. In the last two years, there has been an amount of data supporting a huge improvement about the use of laser therapy for stress incontinence. However, the current insufficient efficacy and safety of laser therapy limits the general recommendations.

3. Pelvic organ prolapse

3.1 Introduction

Pelvic organ prolapse (POP) is a highly prevalent problem in women, defined as the descent of the pelvic organs (i.e., vagina, uterus, bladder, and/or rectum) from their normal anatomical locations into or through the vagina. The prevalence of pelvic organ prolapse is thought to be as high as 30 to 60% of parous women (all stages of prolapse) [56, 57] and will present an increase due to the aging population, as well as the increase prevalence of undergoing future surgical procedures. Most cases of pelvic organ prolapse are asymptomatic. For those with symptomatic prolapse, around 3–6% are reported with a POP-related decisive lower quality of life, including bladder and bowel issues as well as coital dysfunction [58, 59]. In the general female population, the lifetime risk of undergoing a single operation for prolapse is about 12–19%, and by age 80 was 11.1% [4, 6, 60, 61].

3.2 The etiology and risk factors for pelvic organ prolapse

The pelvic organs are supported by the fibromuscular connective tissue in the vaginal wall, pelvic ligaments, and levator ani muscles. Tears in the pelvic floor muscles, fascial tissues, and/or weakness of the ligaments lead to prolapse. Causes of pelvic organ prolapse are complex and multi-factorial. Many risk factors are correlated with POP, including pregnancy, ethnicity or race, operative vaginal childbirth, higher parity, forceps delivery, aging, previous hysterectomy, menopause, connective tissue disorders, chronic constipation, weakness of the pelvic floor, increasing body mass index, and family history of prolapse.

3.3 Clinical manifestation of pelvic organ prolapse

Pelvic organ prolapse encompasses several prevalent clinical conditions, including vaginal bulging symptoms, urinary and fecal incontinence, obstructed defecation, sexual dysfunction, and female pelvic organ prolapse (FPOP). These are closely associated with the site of prolapse (the anterior vaginal wall, posterior vaginal wall, and apical prolapse). A feeling or visibility of vaginal bulging is the most frequently reported and specific symptom of pelvic organ prolapse [62]. Pelvic organ prolapse does not always present any symptoms and is discovered incidentally on physical examination [63]. If the prolapse reaches roughly the level of the hymenal remnant, women can present with vaginal symptoms of prolapse.

The symptoms of pelvic organ prolapse vary by the extent of prolapse and whether the prolapse involves the anterior, posterior, or apical segments of the vagina (see **Box 3**).

Vaginal symptoms

- Sensation or visibility of vaginal bulging or protrusion.
- Pelvic or vaginal pressure/heaviness.

Urinary symptoms

- Symptoms related to bladder storage (urinary incontinence, urgency, frequency)
- Voiding symptoms (hesitancy, intermittency, prolonged urinary steam, and insufficient emptying)
- Splinting (pressing on the prolapse to help empty bladder)
- Position-dependent voiding

Anorectal symptoms

- Incontinence of stool.
- Sensation of anorectal blockage or incomplete emptying.

- Hard straining/urgency to defecate.
- Splinting vagina or perineum for complete evacuation.

Coital symptoms

- Dyspareunia
- Vaginal dryness
- Arousal or orgasm reduction
- Obstructed intercourse

Pain

- Pain in vagina, bladder, low back, or rectum.

Box 3.
The symptoms of pelvic organ prolapse are performed.

3.4 Diagnosis and evaluation of pelvic organ prolapse

Pelvic organ prolapse can be confirmed according to the patient's clinical symptoms and physical examination.

3.4.1 History

A detailed history should be offered. It contains the severity of prolapse-related symptoms, duration, aggravating factors, evaluation of voiding or defecation function, and accompanying coital dysfunction. In addition, based on the including risk factors for prolapse, a complete history of medical, obstetric, and surgical should be conducted [63]. Of note, some of bladder, bowel, and sexual symptoms may not be caused by prolapse but coexist with pelvic organ prolapse, such as overactive bladder, recurrent urinary tract infections, chronic constipation as well as urinary incontinence [64]. The site of prolapse is characterized as the anterior vaginal wall, posterior vaginal wall, and apical prolapse. The most common type of pelvic organ prolapse is the prolapse of the anterior vaginal wall, or cystocele, discovered 2-fold more common than posterior vaginal prolapse (the rectocele), and 3-fold as often as apical prolapse [65, 66]. Anterior-predominant prolapse presents an almost three times more apical descent below normal than posterior-predominant vaginal prolapse [67]. Among them, anterior pelvic prolapse is closely related to urinary system diseases, including cystocele (bladder bulge from the vagina) and urethral prolapse (urethral overactivity). Symptoms related to urinary symptoms include voiding dysfunction, incontinence, urgency, and frequency, and symptoms like outlet obstruction are commonly reported in cases of symptomatic POP [62]. Posterior pelvic prolapse (rectal prolapse) refers to a herniation of the rectal tissue into the vaginal lumen, resulting in a vaginal bulge along the posterior vaginal wall on examination, while apical prolapse is related to a uterine bulge and/or a hernia of the small intestine into the vagina(enterocele).

3.4.2 Physical examination

The commonly used method is the pelvic organ prolapse quantification (POP-Q) system proposed by the International Continence Society in 1996 (**Figure 3**) [68]. The hymen was defined as the fixed anatomic point of reference throughout this descriptive system of quantitative prolapse. There are six defined points above or proximal to the hymen (negative centimeters number) or centimeters below or distal to the hymen (positive centimeters number) and the plane of the hymen is defined as zero. Ordinal stages of pelvic organ prolapse can be assigned into five stages (0 through IV), in accordance with the most severe and distal portion of the prolapse by using specific letter qualifiers, as follows in **Table 1** and **Figure 4**. Besides, there are three different types of prolapse are shown in **Figure 5**.

A speculum is performed to assess the extent of prolapse. It should be noted that more commonly prolapse involves multiple compartments of the vagina in most cases of symptomatic POP [64]. Attention should be paid in evaluating all three compartments during examination. A bivalve speculum combined with a single blade can be performed to assess the anterior and posterior vaginal walls during a Valsalva maneuver, respectively. Once apical prolapse has been visualized, an assessment of occult stress urinary incontinence is allowed during a Valsalva maneuver or cough.

3.5 Management of pelvic organ prolapse

Prolapse is a benign disease. If women with a mild degree of asymptomatic prolapse, observation is commonly recommended. Some women with advanced POP (stage 3 or 4) can also have few symptoms, little or no bother as well. In these cases, observation is considered first [70]. The assessment and treatment are offered to women with POP only if they develop bothersome symptoms due to the prolapse. In

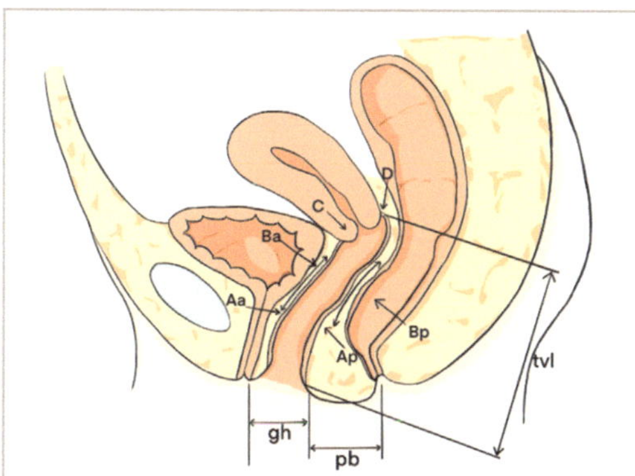

anterior wall Aa	anterior wall Ba	cervix or cuff C
genital hiatus gh	perineal body pb	total vaginal length tvl
posterior wall Ap	posterior wall Bp	posterior fornix D

Figure 3.
The sites of defined points: Aa = a mid-point on the anterior vaginal wall, 3 cm proximal to the external urethral meatus, Ba = represents the most distal position of any part of the upper anterior vaginal wall from the vaginal cuff or anterior vaginal fornix to point Aa, C = the most distal edge of the cervix or cuff (vaginal apex (upper vagina)), D = the site of the posterior fornix (or pouch of Douglas) in a woman, Bp = the most distal site of any part of the upper posterior vaginal wall from the vaginal cuff or posterior vaginal fornix to point Ap, Ap = in the midline of the posterior vaginal wall 3 cm proximal to the hymen. a = anterior vaginal wall, p = posterior vaginal wall, Cx = cervix, gh = genital hiatus, pb = perineal body, tvl = total vaginal length (adapted from Bump et al. [68]).

POP-Q staging	Measurements are in centimeters and measured from the hymen
Stage 0	• No prolapse. Point C or D is within 2 cm of TVL. Aa, Ba, Ap, and Bp are all 3 cm above the hymen (i.e., −3).
Stage 1	• The vaginal apex is lower than 2 cm from TVL. Aa, Ba, Ap, and Bp are higher than −3 cm but less than −1 cm from the hymen (above the hymen).
Stage 2	• Prolapse is within 1 cm of the plane of the hymen (1 ≤ hymen≥ − 1 cm).
Stage 3	• The most distal prolapse is greater than 1 cm below the hymen but less than stage 4 (+ (TVL-2 cm).
Stage 4	• complete vaginal eversion • the maximum prolapse is outside of the body to within 2 cm of the TVL (TVL-2 cm) (e.g., if the TVL is 9 cm, the maximal prolapse is +7).

POP-Q = pelvic organ prolapse quantification; TVL = total vaginal length. Data from Bump et al. [68].

Table 1.
Stages of pelvic organ prolapse by pelvic organ prolapse quantification examination.

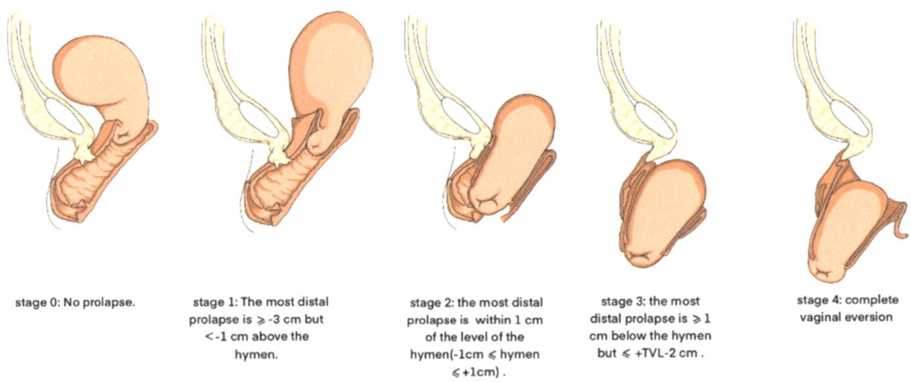

Figure 4.
Stages of prolapse are performed in accordance with the pelvic organ prolapse quantification system (POP-Q) (adapted from Haylen [69]).

brief, treatment selection for women with POP contains observation, conservative measures, pessary use, and surgery.

3.5.1 Conservative management

Generally, the purpose of nonsurgical treatment is to alleviate symptoms, and increase the strength, endurance, and support of pelvic floor muscles. Conservative or mechanical treatments are considered for women with a mild degree of symptomatic prolapse, or those who wish to preserve fertility but are unwilling to accept or cannot tolerate surgery. Pelvic floor muscle training and pessary use are the demonstrated conservative treatment options for patients with POP. Lifestyle modifications with little risk, including weight loss, minimizing straining or constipation, and avoidance of heavy lifting, coughing, or high-impact exercise, are commonly offered.

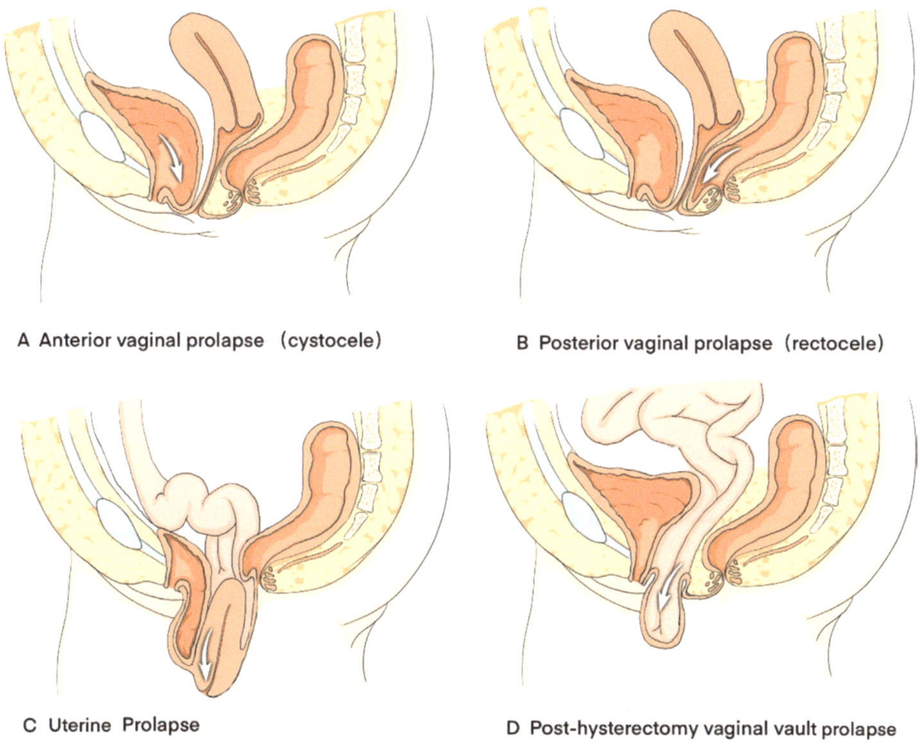

Figure 5.
Different types of pelvic floor prolapse. A = anterior vaginal prolapse, B = posterior vaginal wall prolapse, C = uterine prolapse, D = vaginal vault prolapse (posthysterectomy) (adapted from Haylen [69]).

3.5.1.1 Pelvic floor muscle training

Pelvic floor muscle training, recommended as first-line therapy for POP, refers to systematic contraction of the levator ani muscles with the goal of improving pelvic floor function and symptoms. It is demonstrated pelvic floor muscle training is effective in improving symptoms in women with mild-to-moderate POP [71, 72].

3.5.1.2 Pessary usage

Pessaries are intravaginal mechanical devices inserted into the vagina canal. They can be offered as short- or long-term therapy or hold over the need for surgery, through helping support the prolapsed organs inside the vagina and related pelvic structures, as well as relieving pressure on the bladder and bowel. Pessaries are indicated in patients who wish to childbearing, do not desire surgery, are frail, or have comorbidities that are not medically fit for surgery intervention. Pessaries are available in various shapes and sizes, helping to adapt to differences vaginal diameter and length in anatomy. The most commonly used pessaries are the ring, ring with support, Gellhorn, and doughnut pessaries. A ring pessary or ring with support is the first option for most patients due to an easy insertion and removement. Ring pessaries present a more effective improvement at an early stage of prolapse, and Gellhorn pessaries are usually conducted for patients with more advanced degrees of prolapse.

Pessaries are a relatively simple and effective method for patients with POP but not perfect, and as high as 92% of patients among them can be successfully placed with a pessary [73]. A systematic review, including 89 articles, reported that more than 50% of women continued pessary use and had an obvious improvement in their quality of life, sexual function, and body perception [74]. Complications from pessaries insertion include discontinuation, local infection, vaginal bleeding, the risk of developing vaginal erosions or discharge, and new bowel and bladder symptoms [75, 76]. Although these are typically minor, an appropriate short-term removal or vaginal estrogen therapy should be considered. Moreover, a recent Cochrane review found that pessary use with pelvic floor muscle exercises probably improves prolapse-specific symptoms and quality of life compared with exercises alone [77].

3.5.2 Surgical treatments

Surgery is generally indicated in patients with bothersome symptomatic POP who have at least stage 2 prolapse on examination and have failed or declined nonsurgical therapy [78]. The objective of POP repair is to restore normal anatomy and function of the vagina, lower urinary tract, and pelvic organs, and thus eliminate POP-related symptoms. The option for the approach is based on several factors: the patient's expectations, desires, and needs, degree of prolapse, comorbidities, and surgeon preference and capability. Surgical approaches include reconstructive and obliterative procedures. The former preserves the vagina while the latter closes the vagina. A wide variety of operations can be conducted *via* vaginal, laparoscopy, robot-assisted, or laparotomy for the treatment of patients with prolapse, each of which has an individual risk-benefit profile. The most common and traditional technique is the anterior repair (colporrhaphy) in terms of anterior vaginal wall prolapse, which is frequently offered to combine with other interventions to treat prolapse and/or urinary stress incontinence. The repair procedures of anterior and posterior defects comprised more than 90% of all prolapse operations [4].

3.5.2.1 Vaginal approach

Transvaginal approaches include vaginal hysterectomy, anterior or posterior vaginal wall repair (colporrhaphy), McCall culdoplasty, Manchester repair (amputation of the cervix with uterus suspension to the cardinal ligaments), prespinous and sacrospinous colpopexy, Le Fortes procedure, colpocleisis, sacrospinous ligament fixation (SSLF), uterosacral ligament suspension (ULS), and perineal reconstruction. They can be used for anterior, apical, or posterior prolapse of any stage. It is reported that 80–90% of approaches are implemented through the vaginal approach [79].

3.5.2.2 Laparoscopic, robotic, or abdominal approach (Sacrocolpopexy)

Transabdominal approaches include sacrocolpopexy, hysterectomy, vault suspending and uterosacral ligament plication, and enterocele ligation. Abdominal surgery can be performed through an open incision or a minimally invasive *via* the laparoscope or robot.

3.5.2.3 Apical pelvic organ prolapse

Apical prolapse is sometimes referred to as uterine or cervical prolapse. Uterine preservation (hysteropexy) was a feasible option for well-selected patients with apical pelvic organ prolapse, who may prefer preserve fertility, concern about the risks of hysterectomy, or a negative impact on coital satisfaction and a sense of identity. Hysterectomy has been usually employed for treating women with uterine defects and vaginal fornix suspension should be used to prevent recurrent prolapse as well. If the apical pelvic prolapse women are still present in women after hysterectomy, a uterosacral ligament suspension (ULS) or sacrospinous ligament fixation (SSLF) with transvaginal procedures can be considered. Some contemporary devices have been applied in clinical practice to reduce severe prolapse, but the data on effectiveness remain to be observed. When a hysterectomy is performed and an enterocele is present, iliococcygeus fascia suspension may be used [80]. In addition, after a total hysterectomy, the prolapse of the vaginal cuff may be present, which is also called vaginal vault prolapse (VVP). Sacrocolpopexy(SC) is regarded as the standard treatment for the management of vaginal vault prolapse, which is available for open operation (OSC), as well as minimally invasive approach assisted by laparoscopy (LSC) and robots (RSC) [81]. An updated network meta-analysis comparing efficacy and safety of open, laparoscopic, and robotic-assisted sacrocolpopexy approaches simultaneously indicated a superiority of robot-assisted and laparoscopic sacrocolpopexy over the open procedures with better anatomic outcomes, less estimated blood loss and overall postoperative complications [82]. In elderly and fragile patients with vaginal vault or severe uterine prolapse, colpocleisis is an appropriately recommended option with the benefit of a high success rate (91–100%), an effective improvement, and low risk [83]. It encompasses a uterine-sparing procedure, the LeFort colpocleisis, colpocleisis with hysterectomy, and posthysterectomy vaginal vault colpocleisis. The goal of the colpocleisis procedure is a well-approximated, obliterated vagina, approximately 3 cm in depth and 1 cm in width, which may help to gain well treated results [84].

3.5.2.4 Anterior pelvic organ prolapse

Materials used in surgery of anterior pelvic prolapse repair include native tissue, synthetic mesh, and biological xenograft material. If a patient has anterior vaginal wall prolapse, an anterior vaginal with native tissue repair can be offered. A recent study enrolling 102 patients with over five years of follow-up indicated that hysteropexy with anterior vaginal native tissue repair may be a suitable surgical treatment for anterior compartment prolapse and concomitant stage II uterine prolapse, and the cure rates of subject and object could achieve up to 88% and 86%, respectively [85]. In addition, Y-shaped synthetic polypropylene (PP) mesh was introduced in the middle of the twentieth century. It can be performed in sacrocolpopexy, fixed to the anterior, apical, and posterior aspects of the vagina as well as the anterior longitudinal ligament overlying the sacrum, restoring the normal axial position of the vagina. However, the use of polypropylene mesh has become highly controversial due to reported serious complications involving chronic pain and mesh erosion or exposure [86]. In patients with cystocele prolapse, the use of a standardized, trocar-guided mesh kit led to success in higher short-term rates but accompanied higher rates of complications and postoperative adverse events compared to anterior colporrhaphy [87]. Besides, if patients with vaginal wall prolapse caused by lateral and central defects, anterior colporrhaphy combinates with paravaginal repair can be considered. For patients with severe anterior prolapse, an assessment is recommended to be conducted to determine the risk of concomitant potential SUI before prolapse surgery, which is helpful in formulating surgical plans. In brief, although most materials used in the repair of anterior prolapse may lead to some complications, mesh has potential benefits in some well-selected patients. If surgeons and patients can understand the potential risks and benefits, then vaginal mesh implantation can be considered.

3.5.2.5 Posterior pelvic organ prolapses

The posterior wall prolapse encompasses a variety of bothersome symptoms, including a herniation of the anterior rectal wall (rectocele), a vaginal bulge, constipation, and coital disorder. If the defect extends to the proximal posterior vaginal wall, prolapse could manifest in the form of a herniation of small bowel or an enterocele. Patients with posterior vaginal prolapse sometimes mainly manifested the need for using vaginal splints or fingers to facilitate defecation. The best type of posterior vaginal wall repair lack of enough evidence. Repair procedures should be performed according to the extent of prolapse symptoms and patients' quality of life. Also, the main goal of rectocele repair is to moderate bothersome symptoms, build up anatomic support, and recover defecation and coital function. Rectocele repair can be offered *via* a transvaginal, transanal, abdominal, laparoscopic, and transperineal route. Transvaginal rectocele native tissue repairs contain midline plication and site-specific defect repairs. Some studies indicated that the transvaginal surgical approach was probably superior to the transperineal or transanal route [88, 89].

3.5.2.6 Patients with POP and concurrent SUI

It is reported that POP is often correlated with symptomatic or occult stress urinary incontinence, approximately 55% of women with stage 2 POP have concomitant SUI, and only 33% of women with stage 4 POP have SUI, maybe as the result of kinking of the urethra if the prolapse advances [90]. The occult SUI is discovered only with the use of a speculum or pessary during physical examination. Patients with occult SUI may develop SUI symptoms following POP surgery. In addition, women with POP and no concomitant SUI may be at the risk of developing stress urinary incontinence (De nova SUI) symptoms after POP repair procedures. Therefore, a prolapse reduction cough stress exam should be conducted for preoperative women with POP and the POP operations combined with contemporaneous or delayed continence procedures should be offered to women with POP and SUI (symptomatic or occult). Adding an MUS during vaginal POP repair is recommended to reduce the postoperative SUI. However, the risk of short-term postoperative difficulty in urination void and other lower urinary tract symptoms needs to be considered [91]. It is indicated that the retropubic MUS is associated with greater morbidity than transobturator MUS [42].

3.6 Treatment of pelvic floor reconstruction

The treatment interventions for patients with pelvic floor dysfunction should be adjusted according to their specific needs. A multidisciplinary approach is usually necessary. In the past decade, the use of grafts (absorbable and permanent mesh, biological grafts) repair has been a controversial approach to treat POP, but the use of synthetic and biological transplant materials to improve the integrity and durability of POP repair has gradually become popular in selected cases. Based on current evidence, using vaginal mesh in the pelvic floor is associated with a well-established risk of complications, which has become a debated issue on a global level. Although the majority of females report treatment success and satisfaction, mesh-related complications after pelvic floor reconstruction are an advancing condition with various clinical manifestations, including mesh extrusion/exposure (vaginal/ urinary tract), mesh degradation, urinary tract perforation, vascular injury, de novo voiding LUTS and storage LUTS, vaginal exposure, vaginal infection, dyspareunia, and the need for a second procedure to remove or revise the mesh. In terms of psychological trauma, strained intimate relationships were also affected, with reduced sexual functioning and intimacy, as well as emotional suffering, which included anxiety and fears related to suicidal thoughts [92–95]. The mesh problem, extrusion or erosion location, pelvic floor defect extent, severity, and probable recurrence should all be taken into consideration when choosing the appropriate treatment plan. With a developed understanding of the anatomy and function of the pelvic floor, pelvic floor reconstruction has received much attention and more new technologies are also constantly emerging, developed to improve the traditional methods. Robotic surgery has gained popularity for the reconstruction of pelvic floor defects. Compared with traditional transvaginal or transperineal procedures, robotic surgery is a probable safe and effective indication for pelvic floor reconstructive surgery in terms of better functional outcomes and long-term successful rates for the treatment of advanced or multicompartmental prolapses. The ultimate target of pelvic floor reconstruction is to restore the normal anatomical structure and function of the vagina, bladder, and surrounding tissues. A correct evaluation of the anatomical structure and function of the pelvic floor is crucial for the development of new therapeutics and optimization on prognosis. In addition, as people increasingly value the quality of life, the cost of healthcare is also increasing. Therefore, people are still making continuous efforts to explore methods for quantifying symptoms and evaluating outcomes, and some safe and more effective new technologies are still emerging to correct pelvic floor disorders, alleviate symptoms, and improve quality of life.

Author details

Yueying Chen and Ran Pang*
Urology Department, Guang'anmen Hospital, China Academy of Chinese Medical Sciences, Beijing, China

*Address all correspondence to: pangran2002@sina.com

References

[1] Hagen S, Stark D. Conservative prevention and management of pelvic organ prolapse in women. The Cochrane Database of Systematic Reviews. 2011; **12**:CD003882

[2] Nygaard I, Barber MD, Burgio KL, et al. Prevalence of symptomatic pelvic floor disorders in US women. JAMA: The Journal of the American Medical Association. 2008;**300**(11):1311-1316

[3] Smith CP. Male chronic pelvic pain: An update. Indian Journal of Urology: IJU: Journal of the Urological Society of India. 2016;**32**(1):34-39

[4] Olsen AL, Smith VJ, Bergstrom JO, et al. Epidemiology of surgically managed pelvic organ prolapse and urinary incontinence. Obstetrics and Gynecology, United States. 1997;**89**(4):501-506

[5] Fialkow MF, Newton KM, Lentz GM, et al. Lifetime risk of surgical management for pelvic organ prolapse or urinary incontinence. International Urogynecology Journal and Pelvic Floor Dysfunction. 2008;**19**(3):437-440

[6] Fj S, Cd H, Re M, et al. Lifetime risk of undergoing surgery for pelvic organ prolapse. Obstetrics and Gynecology. 2010;**116**(5):1096-1100

[7] Wu JM, Vaughan CP, Goode PS, et al. Prevalence and trends of symptomatic pelvic floor disorders in U.S. women. Obstetrics and Gynecology. 2014;**123**(1): 141-148

[8] Petros PE, Ulmsten UI. An integral theory and its method for the diagnosis and management of female urinary incontinence. Scandinavian journal of Urology and Nephrology. Supplementum, England. 1993; **153**:1-93

[9] Quintana Franco LM, González López R, Garde García H, et al. Evolution and current status of the management of functional and pelvic floor pathology in the hospitals of the Community of Madrid. Actas Urológicas Españolas (English Edition). 2023;**47**(3):187-192

[10] Faubion S, Shuster LT, Bharucha AE. Recognition and management of nonrelaxing pelvic floor dysfunction. Mayo Clinic Proceedings. 2012;**87**(2): 187-193

[11] Rehder P, Staudacher NM, Schachtner J, et al. Hypothesis that urethral bulb (corpus spongiosum) plays an active role in male urinary continence. Advances in Urology. 2016; **2016**:6054730

[12] DeLancey JO. Structural aspects of the extrinsic continence mechanism. Obstetrics and Gynecology. 1988;**72**(3 Pt 1):296-301

[13] Brooks JD, Chao WM, Kerr J. Male pelvic anatomy reconstructed from the visible human data set. The Journal of Urology. 1998;**159**(3):868-872

[14] Gadda F, Carmignani L, Favini P, et al. [Anatomy of the urethral sphincteric vesico-prostatic complex]. Archivio Italiano Di Urologia, Andrologia: Organo Ufficiale [di]. Societa Italiana Di Ecografia Urologica E Nefrologica. 2001; **73**(3):115-117

[15] Walz J, Burnett AL, Costello AJ, et al. A critical analysis of the current knowledge of surgical anatomy related to optimization of cancer control and preservation of continence and erection in candidates for radical prostatectomy. European Urology. 2010;**57**(2):179-192

[16] Twiss C, Triaca V, Rodríguez LV. Familial transmission of urogenital prolapse and incontinence. Current Opinion in Obstetrics & Gynecology. 2007;**19**(5):464-468

[17] Bergman A, Bhatia N. Urodynamic appraisal of the Marshall-Marchetti test in women with stress urinary incontinence. Urology. 1987;**29**(4):458-462

[18] Walters MD, Diaz K. Q-tip test: A study of continent and incontinent women. Obstetrics and Gynecology. 1987;**70**(2):208-211

[19] Dmochowski R, Blaivas JM, Gormley EA, et al. Update of AUA guideline on the surgical management of female stress urinary incontinence. The Journal of Urology. 2010;**183**(5):1906-1914

[20] Nager CW, Brubaker L, Litman HJ, et al. A randomized trial of urodynamic testing before stress-incontinence surgery. The New England Journal of Medicine. 2012;**366**(21):1987-1997

[21] Good M, Solomon ER. Pelvic floor disorders. Obstetrics and Gynecology Clinics of North America. 2019;**46**(3): 527-540

[22] Wehbe SA, Fariello JY, Whitmore K. Minimally invasive therapies for chronic pelvic pain syndrome. Current Urology Reports. 2010;**11**(4):276-285

[23] Desrosiers L, Knoepp LR. Botulinum toxin a: A review of potential uses in treatment of female urogenital and pelvic floor disorders. The Ochsner Journal, Ochsner Clinic, L.L.C. and Alton Ochsner Medical Foundation. 2020;**20**(4):400

[24] Emel G, Nurcan Ç. The development and effectiveness of a care protocol using the stevens star model of knowledge transformation in female patients with stress incontinence: An experimental study. Wound Management & Prevention. 2021;**67**(3):36-47

[25] Lukanović D, Blaganje M, Rhazi I, et al. Urethral Bulkamid® injection after failed midurethral sling: A step-by-step video. International Urogynecology Journal. 2023;**34**(11):2843-2845

[26] Oliver R, Thakar R, Sultan AH, et al. Urogynecology triage clinic: A model of healthcare delivery. International Urogynecology Journal and Pelvic Floor Dysfunction. 2009;**20**(8):913-917

[27] Jimènez Torres M, Beitl K, Hummel Jimènez J, Mayer H, Zehetmayer S, Umek W, et al. Benefit of a nurse-led telephone-based intervention prior to the first urogynecology outpatient visit: A randomized-controlled trial. International Urogynecology Journal. 2021;**32**(6):1489-1495

[28] Franken MG, Ramos IC, Los J, et al. The increasing importance of a continence nurse specialist to improve outcomes and save costs of urinary incontinence care: An analysis of future policy scenarios. BMC Family Practice. 2018;**19**(1):31

[29] Ebbesen MH, Hunskaar S, Rortveit G, et al. Prevalence, incidence and remission of urinary incontinence in women: Longitudinal data from the Norwegian HUNT study (EPINCONT). BMC Urology. 2013;**13**:27

[30] Park HK, Chang S, Palmer MH, et al. Assessment of the impact of male urinary incontinence on health-related quality of life: A population based study. Lower Urinary Tract Symptoms. 2015;7(1):22-26

[31] Helfand BT, Smith AR, Lai H, et al. Prevalence and characteristics of urinary

incontinence in a treatment seeking male prospective cohort: Results from the LURN study. The Journal of Urology. 2018;**200**(2):397-404

[32] González-Isaza P, Sánchez-Borrego R, Lugo Salcedo F, et al. Pulsed magnetic stimulation for stress urinary incontinence and its impact on sexuality and health. Medicina (Kaunas, Lithuania), Switzerland. 2022;**58**(12): 1721

[33] Das AK, Kucherov V, Glick L, et al. Male urinary incontinence after prostate disease treatment. The Canadian Journal of Urology. 2020;**27**(S3):36-43

[34] Canning A, Raison N, Aydin A, et al. A systematic review of treatment options for post-prostatectomy incontinence. World Journal of Urology. 2022;**40**(11):2617-2626

[35] Naumann G, Aigmüller T, Bader W, et al. Diagnosis and therapy of female urinary incontinence. Guideline of the DGGG, OEGGG and SGGG (S2k-level, AWMF registry No. 015/091, January 2022). Geburtshilfe und Frauenheilkunde. 2023;**83**(4):377-409

[36] Wu X, Zheng X, Yi X, Lai P, Lan Y. Electromyographic biofeedback for stress urinary incontinence or pelvic floor dysfunction in women: A systematic review and meta-analysis. Advances in Therapy. 2021;**38**(8):4163-4177

[37] Cody JD, Jacobs ML, Richardson K, et al. Oestrogen therapy for urinary incontinence in post-menopausal women. The Cochrane Database of Systematic Reviews. 2012;**10**(10): CD001405

[38] Weber MA, Kleijn MH, Langendam M, et al. Local oestrogen for pelvic floor disorders: A systematic review. PloS One, United States. 2015;**10**(9):e0136265

[39] Goessens EMV, Cammu H. A 10- to 20-year follow-up after tension-free vaginal tape for stress urinary incontinence. International Urogynecology Journal. 2023;**34**(9): 2107-2114

[40] Fusco F, Abdel-Fattah M, Chapple CR, et al. Updated systematic review and meta-analysis of the comparative data on Colposuspensions, pubovaginal slings, and Midurethral tapes in the surgical treatment of female stress urinary incontinence. European Urology, Switzerland. 2017;**72**(4):567-591

[41] Imamura M, Hudson J, Wallace SA, et al. Surgical interventions for women with stress urinary incontinence: Systematic review and network meta-analysis of randomised controlled trials. BMJ (Clinical Research ed.). 2019;**365**: l1842

[42] Kim A, Kim MS, Park Y-J, et al. Retropubic versus Transobturator mid urethral slings in patients at high risk for recurrent stress incontinence: A systematic review and meta-analysis. The Journal of Urology. 2019;**202**(1): 132-142

[43] Mathieson R, Kippen R, Manning T, et al. Stress urinary incontinence in the mesh complication era: Current Australian trends. BJU International. 2021;**128**(1):95-102

[44] Zacche M, Mukhopadhyay S, Giarenis I. Changing surgical trends for female stress urinary incontinence in England. International Urogynecology Journal. 2019;**30**(2):203-209

[45] Wyndaele M, Madhu C, Hashim H. Female stress urinary incontinence surgery: 'Resurgence of the titans'.

Turkish Journal of Urology. 2021;**47**(3): 210-215

[46] Zheng Y, Rovner E. Update on urethral bulking for stress urinary incontinence in women. Current Urology Reports. 2022;**23**(10): 203-209

[47] Serati M, Braga A, Salvatore S, et al. Up-to-date procedures in female stress urinary incontinence surgery: A concise review on bulking agents procedures. Medicina, Multidisciplinary Digital Publishing Institute (MDPI). 2022; **58**(6):775

[48] Serati M, Braga A, Salvatore S, et al. Up-to-date procedures in female stress urinary incontinence surgery: A concise review on bulking agents procedures. Medicina (Kaunas, Lithuania). 2022;**58** (6):775

[49] Chouhan JD, Terlecki RP. A user's guide for surgery involving the artificial urinary sphincter. Sexual Medicine Reviews. 2019;**7**(1):167-177

[50] Lee H-Y, Kuo H-C. Intravesical injection of botulinum toxin type a in men without bladder outlet obstruction and post-Deobstructive prostate surgery. Toxins. 2023;**15**(3):221

[51] Kulseng-Hanssen S, Husby H, Schiøtz HA. Follow-up of TVT operations in 1,113 women with mixed urinary incontinence at 7 and 38 months. International Urogynecology Journal and Pelvic Floor Dysfunction, England. 2008;**19**(3):391-396

[52] Rezapour M, Ulmsten U. Tension-free vaginal tape (TVT) in women with mixed urinary incontinence–a long-term follow-up. International Urogynecology Journal and Pelvic Floor Dysfunction. 2001;**12**(Suppl 2):S15-S18

[53] Sung VW, Borello-France D, Newman DK, et al. Effect of Behavioral and pelvic floor muscle therapy combined with surgery vs surgery alone on incontinence symptoms among women with mixed urinary incontinence: The ESTEEM randomized clinical trial. JAMA. 2019;**322**(11): 1066-1076

[54] Liu X, Li T, Zhang J, et al. Mesenchymal stem cell-based therapy for female stress urinary incontinence. Frontiers in Cell and Developmental Biology. 2023;**11**:1007703

[55] Fang J, Peng T, Liu J, et al. Muscle-derived stem cells combined with nerve growth factor transplantation in the treatment of stress urinary incontinence. Urology. 2022;**166**:126-132

[56] Flynn BJ, Webster GD. Surgical management of the apical vaginal defect. Current Opinion in Urology. 2002;**12**(4): 353-358

[57] Maher C, Feiner B, Baessler K, et al. Surgery for women with anterior compartment prolapse. The Cochrane Database of Systematic Reviews. 2016; **2016**(11):CD004014

[58] Mao M, Ai F, Zhang Y, et al. Changes in the symptoms and quality of life of women with symptomatic pelvic organ prolapse fitted with a ring with support pessary. Maturitas. 2018;**117**:51-56

[59] Verbeek M, Hayward L. Pelvic floor dysfunction and its effect on quality of sexual life. Sexual Medicine Reviews. 2019;**7**(4):559-564

[60] Kurkijärvi K, Aaltonen R, Gissler M, et al. Pelvic organ prolapse surgery in Finland from 1987 to 2009: A national register based study. European Journal of Obstetrics, Gynecology, and Reproductive Biology. 2017;**214**:71-77

[61] Løwenstein E, Ottesen B, Gimbel H. Incidence and lifetime risk of pelvic organ prolapse surgery in Denmark from 1977 to 2009. International Urogynecology Journal. 2015;**26**(1): 49-55

[62] Haylen BT, Maher CF, Barber MD, et al. An international Urogynecological association (IUGA)/international continence society (ICS) joint report on the terminology for female pelvic organ prolapse (POP). International Urogynecology Journal. 2016;**27**(4): 655-684

[63] Swift SE, Tate SB, Nicholas J. Correlation of symptoms with degree of pelvic organ support in a general population of women: What is pelvic organ prolapse? American Journal of Obstetrics and Gynecology. 2003;**189**(2): 372-377; discussion 377-379

[64] Raju R, Linder BJ. Evaluation and Management of Pelvic Organ Prolapse. Mayo Clinic Proceedings, Elsevier. 2021; **96**(12):3122-3129

[65] Hendrix SL, Clark A, Nygaard I, Aragaki A, Barnabei V, McTiernan A. Pelvic organ prolapse in the women's health initiative: Gravity and gravidity. American Journal of Obstetrics and Gynecology. 2002;**186**(6):1160-1166

[66] Handa VL, Garrett E, Hendrix S, Gold E, Robbins J. Progression and remission of pelvic organ prolapse: A longitudinal study of menopausal women. American Journal of Obstetrics and Gynecology. 2004;**190**(1):27-32

[67] Berger MB, Kolenic GE, Fenner DE, et al. Structural, functional, and symptomatic differences between women with rectocele versus cystocele and normal support. American Journal of Obstetrics and Gynecology. 2018;**218**(5): 510.e1-510.e8

[68] Bump RC, Mattiasson A, Bø K, et al. The standardization of terminology of female pelvic organ prolapse and pelvic floor dysfunction. American Journal of Obstetrics and Gynecology. 1996;**175**(1): 10-17

[69] Haylen BT, Maher CF, Barber MD, et al. An international Urogynecological association (IUGA) / international continence society (ICS) joint report on the terminology for female pelvic organ prolapse (POP). International Urogynecology Journal. 2016;**27**(2): 165-194

[70] Jelovsek JE, Maher C, Barber MD. Pelvic organ prolapse. Lancet (London, England). 2007;**369**(9566):1027-1038

[71] Hagen S, Stark D, Glazener C, et al. Individualised pelvic floor muscle training in women with pelvic organ prolapse (POPPY): A multicentre randomised controlled trial. Lancet (London, England). 2014;**383**(9919):796-806

[72] Wiegersma M, Panman CMCR, Kollen BJ, et al. Effect of pelvic floor muscle training compared with watchful waiting in older women with symptomatic mild pelvic organ prolapse: Randomised controlled trial in primary care. BMJ (Clinical Research ed.). 2014; **349**:g7378

[73] Cundiff GW, Amundsen CL, Bent AE, et al. The PESSRI study: Symptom relief outcomes of a randomized crossover trial of the ring and Gellhorn pessaries. American Journal of Obstetrics and Gynecology. 2007;**196**(4):405.e1-405.e8

[74] de Albuquerque CSC, de Castro EB, Juliato CRT. Female pelvic organ prolapse using pessaries: Systematic review. International Urogynecology Journal. 2016;**27**(12):1797-1803

[75] Sung VW, Jeppson P, Madsen A. Nonoperative Management of Pelvic Organ Prolapse. Obstetrics and Gynecology. 2023;**141**(4):724-736

[76] Wu V, Farrell SA, Baskett TF, et al. A simplified protocol for pessary management. Obstetrics and Gynecology. 1997;**90**(6):990-994

[77] Bugge C, Adams EJ, Gopinath D, Stewart F, Dembinsky M, Sobiesuo P, et al. Pessaries (mechanical devices) for managing pelvic organ prolapse in women. The Cochrane Database of Systematic Reviews. 2020;**11**(11): CD004010

[78] Barber MD. Pelvic organ prolapse. BMJ (Clinical Research ed.). 2016;**354**: i3853

[79] Serati M, Braga A, Bogani G, et al. Iliococcygeus fixation for the treatment of apical vaginal prolapse: Efficacy and safety at 5 years of follow-up. International Urogynecology Journal. 2015;**26**(7):1007-1012

[80] Meeks GR, Washburne JF, McGehee RP, et al. Repair of vaginal vault prolapse by suspension of the vagina to iliococcygeus (prespinous) fascia. American Journal of Obstetrics and Gynecology. 1994;**171**(6):1444-1452; discussion 1452-1454

[81] Barber MD, Maher C. Apical prolapse. International Urogynecology Journal. 2013;**24**(11):1815-1833

[82] Chang C-L, Chen C-H, Yang SS-D, et al. An updated systematic review and network meta-analysis comparing open, laparoscopic and robotic-assisted sacrocolpopexy for managing pelvic organ prolapse. Journal of Robotic Surgery. 2022;**16**(5): 1037-1045

[83] FitzGerald MP, Richter HE, Siddique S, et al. Colpocleisis: A review. International Urogynecology Journal and Pelvic Floor Dysfunction. 2006;**17**(3): 261-271

[84] Welch EK, Dengler KL, Wheat JE, et al. Colpocleisis techniques: An open-and-shut case for advanced pelvic organ prolapse. Urology. 2023;**S0090-4295** (23):00261-00263

[85] Serati M, Salvatore S, Torella M, et al. Hysteropexy and anterior vaginal native tissue repair in women with anterior and central compartment prolapse: A long term follow-up. Journal of Clinical Medicine. 2023;**12**(7):2548

[86] Seifalian A, Basma Z, Digesu A, et al. Polypropylene pelvic mesh: What went wrong and what will Be of the future? Biomedicine. 2023;**11**(3):741

[87] Altman D, Väyrynen T, Engh ME, et al. Anterior colporrhaphy versus transvaginal mesh for pelvic-organ prolapse. The New England Journal of Medicine. 2011;**364**(19):1826-1836

[88] Balata M, Elgendy H, Emile SH, et al. Functional outcome and sexual-related quality of life after Transperineal versus transvaginal repair of anterior rectocele: A randomized clinical trial. Diseases of the Colon and Rectum. 2020;**63**(4):527-537

[89] Doumouchtsis SK, Raheem A, Milhem Haddad J, et al. An update of a former FIGO working group report on Management of Posterior Compartment Prolapse. International Journal of Gynaecology and Obstetrics: The Official Organ of the International Federation of Gynaecology and Obstetrics. 2020;**148** (2):135-144

[90] Slieker-ten Hove MCP, Pool-Goudzwaard AL, Eijkemans MJC, et al.

The prevalence of pelvic organ prolapse symptoms and signs and their relation with bladder and bowel disorders in a general female population. International Urogynecology Journal and Pelvic Floor Dysfunction, England. 2009;**20**(9):1037-1045

[91] van der Ploeg JM, van der Steen A, Oude Rengerink K, et al. Prolapse surgery with or without stress incontinence surgery for pelvic organ prolapse: A systematic review and meta-analysis of randomised trials. BJOG: An International Journal of Obstetrics and Gynaecology. 2014;**121**(5):537-547

[92] Farag F, Osman NI, Pang KH, et al. Complications of synthetic Midurethral slings: Is there a relevant discrepancy between observational data and clinical trials? European Urology Focus. 2023: S2405-4569(23)00244-4

[93] Iakovlev V, Guelcher SA, Bendavid R. Degradation of polypropylene in vivo: A microscopic analysis of meshes explanted from patients. Journal of Biomedical Materials Research. Part B, Applied Biomaterials. 2017;**105**(2):237-248

[94] Hansen BL, Dunn GE, Norton P, et al. Long-term follow-up of treatment for synthetic mesh complications. Female Pelvic Medicine & Reconstructive Surgery. 2014;**20**(3):126-130

[95] Dibb B, Woodgate F, Taylor L. When things go wrong: Experiences of vaginal mesh complications. International Urogynecology Journal, Springer. 2023;**34**(7):1575

Chapter 4

Urinary Fistulae: Etiology, Diagnosis and Management

Jacob Rubin and Teresa L. Danforth

Abstract

This chapter covers the most common and clinically relevant types of urinary fistulae. Fistulae are classified by the involved organs and discussed separately. For each type of fistula, its pathogenesis, clinical presentation, recommended workup and management options are reviewed. Advantages and disadvantages of different surgical approaches are explored, however detailed instruction on surgical techniques is not provided. This chapter should prepare the clinician to recognize risk factors and symptoms of urinary fistula, initiate the appropriate diagnostic steps, and guide patients toward informed decisions between the available therapeutic options.

Keywords: vesicovaginal, ureterovaginal, urethrovaginal, vesicouterine, colovesical, ureteroarterial

1. Introduction

A fistula is an abnormal passage between a hollow organ and the body surface, or between two hollow organs. Urinary fistulae are those that involve any part of the urinary tract—kidneys, ureters, bladder or urethra. Most urinary fistulae in the industrialized world are iatrogenic and are caused by surgery, but they may also be caused by malignancy, radiation, inflammation, prolonged childbirth and a variety of other pathologies. The resultant leakage of urine often leads to great psychological distress and social isolation in afflicted patients. Although some fistulae heal with only conservative management, the vast majority require definitive surgical repair in the operating room. Surgical techniques for the repair of urinary fistula have evolved over time, their merits intensely debated in the medical literature such that the optimal approach for a given clinical scenario is often difficult to ascertain. Therefore, patients should be evaluated on a case-by-case basis and selection of treatment should take into account location and etiology of the fistula, patient goals of care and functional status, and surgeon experience and preference.

IntechOpen

2. Vesicovaginal fistula

2.1 Etiology

Vesicovaginal fistula (VVF) is an abnormal connection between the bladder and vagina that allows continuous, involuntary leakage of urine into and out of the vagina. The incidence of VVF has been estimated at 0.3 to 2% [1]. In developing countries, the predominant cause of VVF (97%) is prolonged obstructed labor [2]. It has been estimated there are between two and three million women with untreated fistula worldwide, with >95% of obstetric etiology and in developing nations [2]. The presenting fetal part presses against the pelvic floor causing extensive ischemic injury, subsequent soft tissue necrosis and fistula formation. In such cases, fistula is often only one component of devastating multifaceted pathology referred to as *obstructed labor injury complex*, which includes rectal atresia, incompetence of the anal sphincter, destruction of the urethra or cervix, infertility, renal failure, and nerve injury [3]. Treatment of obstetric fistulae are hindered by poor access to care in endemic regions.

In the developed world, advances in obstetric care have reduced the incidence of obstetric fistulae dramatically, and the vast majority of VVF are iatrogenic. The most common cause is gynecological surgery, principally hysterectomy [2], though pelvic malignancy and radiation are also significant causes. In a series of 348 cases of urinary fistula in the UK from 2011, over half of the cases resulted from hysterectomy [4]. Risk of post-hysterectomy fistula formation appears to be influenced considerably by surgical approach and indication. In a report on 343,771 women undergoing hysterectomy in the English National Health Service between 2000 and 2008, the lowest rate of fistula formation was observed after vaginal hysterectomy for prolapse (1 in 3861), and the highest rate was observed after radical hysterectomy for cervical cancer (1 in 87) [5]. VVF involve the posterior surface of the bladder abutting the reproductive tract. A systematic review of 15 studies found that 55.7% of VVF involved the supra-trigonal posterior bladder wall, 27.54% involved the trigone, and the remainder had varied descriptions including "juxta-cervical" [6].

2.2 Presentation and workup

Post-hysterectomy patients with VVF typically present with continuous urinary incontinence, and in the majority of cases onset is delayed between 3 days and 8 weeks postoperatively, suggesting that fistula formation is less often the result of direct operative injury and more often the result of compromised blood supply from thermal cautery or dissection and delayed tissue necrosis [4]. VVF resulting from pelvic radiation may not present for years following completion of therapy due to the slow progression of the underlying obliterative endarteritis. Careful patient history helps to differentiate VVF from other causes of incontinence such as stress, urgency and overflow incontinence. A pelvic exam should always be performed to assess for masses or other concurrent pelvic pathology. At this time a simple dye test may be performed wherein a gauze roll is inserted into the vagina and methylene blue or other dye instilled into the bladder per urethra. After several minutes the gauze is removed and examined for staining. Alternatively, cystography or CT cystography can be performed, however in both methods a small fistula may evade detection if the bladder is not distended adequately to force the dye or contrast agent through the fistula. Cystoscopy should be performed to delineate any concurrent bladder pathology and to attempt to localize the fistula, especially in relation to the ureteral orifices as ureteral reimplantation may be required if the ureter(s) are involved. The most common location of the injury is at the vaginal cuff in the supratrigonal posterior bladder wall [1]. Retrograde pyelogram performed at the time of cystoscopy or CT urogram can be used to evaluate for ureteral involvement, which is present in up to 12% of cases. Cross-sectional imaging of the pelvis is recommended if malignancy is suspected.

2.3 Management

2.3.1 Conservative management

An initial trial of catheter drainage with anticholinergic or beta-3 agonist medications is warranted in select patients with small, uncomplicated fistulae (<1 cm) and no history of pelvic radiation. Conservative management is most likely to be successful when the patient presents early on before the fistula has fully epithelialized, after which it is unlikely to heal spontaneously. However, patients should be forewarned that surgery is likely to become necessary. A meta-analysis of 124 studies reported that of 239 women managed with catheter drainage for 2–12 weeks, only 8% had spontaneous resolution of the fistula and the rest ultimately underwent surgical repair [2].

2.3.2 Early vs. late repair

Perhaps the most controversial aspect of VVF repair is the ideal timing of surgery. There is no good consensus on the definitions of early and late repair, with "early" ranging broadly from 1 to 3 months in the literature and "late" ranging from 2 to 4 months, and these intervals overlapping across studies [7]. One faction of experts advocates for early repair, drawing attention to the diminished quality of life of women awaiting surgery, while other groups advocate delaying repair for at least 6–8 weeks to allow for resolution of inflammation and edema as well as demarcation and sloughing of necrotic tissue [1, 7]. Yet other studies report no difference in outcomes. In a meta-analysis by Bodner et al., no statistically significant difference could be detected in the success rate between 241 patients undergoing repair <12 weeks from fistula occurrence to repair, and 223 patients undergoing repair >12 weeks [2]. In general, it is encouraged to repair uncomplicated postsurgical fistula as soon as possible in order to minimize patient suffering. Notable exceptions are post-obstetric and radiation fistulae, in which cases a waiting period of 3 to 6 months or longer may be appropriate.

2.3.3 Vaginal vs. abdominal approach

Another contentious aspect of VVF repair is the choice between vaginal and abdominal (open, laparoscopic or robotic) approaches. In reality, this decision is largely dictated by the surgeon's experience, and the best approach is likely the one that he or she is most comfortable with. There are no randomized trials comparing the two approaches, but there is substantial data from multiple retrospective studies indicating that both approaches offer excellent outcomes with success rates typically exceeding 90% and approaching 100% in some series. The transvaginal approach is associated with less morbidity and shorter hospital stay and is therefore preferred in most cases, with the abdominal approach reserved for cases where there is concurrent pelvic pathology requiring additional procedures, such as ureteric involvement requiring reimplantation or diminished bladder capacity requiring augmentation cystoplasty [1, 2, 4, 7]. Traditionally the fistulous tract is excised during the repair, but some have criticized this step for causing increasing bleeding and creating an unnecessarily large defect [1, 7].

2.3.4 Interposition grafts

In most cases, an uncomplicated VVF requires only a multilayer, tension-free repair. In a series of 207 patients that underwent VVF repair, the success rate for fistulas <2 cm without tissue interposition was 99% after the first attempt [1]. When tissue quality is poor due to radiation or previous surgery, interposition grafts are used to provide a well-vascularized barrier between bladder and vaginal suture lines and reduce the risk of recurrence. Commonly omentum is used in the abdominal approach, and a Martius flap of fibrofatty tissue from the labia is used for vaginal approach. In cases of a high (proximal) VVF it may be difficult to mobilize a Martius flap adequately, and in these cases gracilis muscle or peritoneal flaps may be used instead [7]. Use of a peritoneal flap, which is readily accessible from the proximal vaginal vault, has been shown to result in comparable outcomes to use of the Martius flap without the need for a second skin incision [1]. Some studies detected no statistically significant difference in success rates based on use of interposition grafts; nonetheless, some surgeons prefer to incorporate these grafts routinely.

2.3.5 Post-obstetric fistula repair

It is worth emphasizing the difference in pathophysiology between post-obstetric fistulae, which are surrounded by large areas of de-vascularized tissue, and postsurgical fistulae which are typically surrounded by healthier tissue. The management of post-obstetric fistulae thus presents a more complex surgical challenge. First, repair of post-obstetric VVF necessitates the use of tissue flaps with adequate vascular pedicle to bring new blood supply to the surgical site in virtually all cases [3]. Second, total loss of the urethra occurs in about 5% of cases [3] and is complicated by a lack of surrounding viable tissue with which to reconstruct a neourethra. Third, urinary incontinence persists even after successful fistula closure in a significant portion of post-obstetric fistula patients [3] and may be due to damage to the bladder neck, urinary sphincter or the pudendal branches that provide its innervation. Fourth, rectovaginal fistula may occur concurrently with VVF, and the rate of successful combined repair has been reported to be equivalent to that of isolated rectovaginal fistula (78%) [3], which is comparatively lower than that of VVF.

2.3.6 Postoperative considerations

Postoperatively, continuous urinary drainage with suprapubic and/or urethral catheterization is paramount to prevent distention of the bladder and tension and breakdown of suture lines. A voiding cystogram is recommended at 2–3 weeks to rule out extravasation or leak, at this time contrast and/or methylene blue is instilled into the catheter and a tampon or vaginal packing is examined for staining to confirm closure of the fistula.

3. Ureterovaginal fistulae

3.1 Etiology

Ureterovaginal fistula (UVF), like VVF, most often occurs as a result of injury during gynecologic surgery, though it may also arise after vaginal delivery, infertility treatment involving oocyte retrieval, pelvic radiation and pessary use [8]. Hysterectomy for a benign indication is most often the offending procedure, with the incidence estimated at 1 UVF per 250 hysterectomies [8]. Furthermore, total abdominal, total laparoscopic and robotic hysterectomy are more likely to result in UVF than supracervical hysterectomy [8].

3.2 Presentation and workup

Patients present with a chief complaint of continuous urinary leakage per vagina. The normal urge to void may differentiate UVF from VVF, as in cases of UVF the uninjured ureter continues to fill the bladder normally, whereas in VVF the bladder is continuously drained via the fistula; however, it should be remembered that VVF and UVF may occur concurrently. The recommended initial workup is a dual-dye tampon test, which is inexpensive and may be performed in the outpatient setting. Patients are prescribed oral phenazopyridine to take before the visit. In the office, a gauze roll is placed in the vagina and methylene blue dye is instilled into the bladder via a Foley catheter. After approximately 30 minutes the gauze roll is removed and examined for presence of orange staining (indicates UVF) and/or blue staining (indicates VVF).

3.3 Management

3.3.1 Primary management

Historically, UVF were managed with ureteroneocystostomy (ureteral reimplantation), urinary diversion with percutaneous nephrostomy, or even nephrectomy [8–10]. However, more recent studies have reported that endoscopic ureteral stenting results in high cure rates for post-hysterectomy UVF and should be considered first-line therapy [8, 9]. Cystoscopy with retrograde pyelography is first used to confirm the diagnosis and localize the fistula, as well as to rule out concurrent VVF. Retrograde stent placement may then be attempted at the same time as retrograde pyelogram. If retrograde stent placement is unsuccessful, antegrade stent placement may be attempted by an interventional radiologist. Rates of successful stent placement vary widely in the literature from 5–92% in different series [8]. The optimal length of ureteral stenting is unclear, with studies reporting protocols from 4 to 8 weeks [9, 10] to 3 months [8]. Rates of resolution of UVF after successful stent placement have been reported from 32% [11] to 100% [8, 9]. A meta-analysis of 17 retrospective studies found that for 285 patients undergoing endoscopic stent placement for UVF, the overall pooled success rate was 32%. However, when stratified by time from fistula diagnosis to intervention, it was shown that success rates were 95% if stenting was performed within 2 weeks, 63% within 6 weeks and 20% after 6 weeks [11]. This suggests that the success of endoscopic stent placement as the sole treatment modality for UVF decreases with time from diagnosis, and that stent placement could potentially obviate the need for more invasive surgery if performed early enough. In contrast, success rate of surgical intervention approached 100% regardless of whether or not it was performed within 6 weeks from diagnosis [11]. Management with urinary diversion with percutaneous nephrostomy is not effective [10]. Several criteria for primary surgical intervention with ureteral reimplantation have been suggested including concurrent VVF, history of ureteral injury with repair at the time of index surgery, and extremely long delay from index surgery to diagnosis [8].

3.3.2 Post-operative considerations

Ureteral strictures have been reported in 6–38% of UVF managed conservatively and may have a delayed presentation. For this reason, annual follow-up to rule out stricture formation and urinary obstruction is recommended. Renal ultrasound, CT urogram, Mag-3 Lasix renogram and retrograde pyelogram may be used as surveillance studies.

4. Urethrovaginal fistula

4.1 Etiology

Urethrovaginal fistula is most often caused by operations of the vagina and urethra including urethral diverticulectomy, urethral sling procedures and anterior colporrhaphy. Excision of urethral diverticulum is an uncommon cause of urinary fistula overall, but the most common cause of urethrovaginal fistulae, accounting for 39.5% of all urethrovaginal fistulae [4]. Other causes include traumatic pelvic fractures, obstetric trauma, vaginal and urethral malignancies, pelvic radiation, and urethral erosion from chronic catheterization or pessary use [12].

4.2 Presentation and workup

Presentation depends on the size and location of the fistula. Large or proximal urethrovaginal fistulae may cause stress incontinence or continuous incontinence akin to VVF. If the fistula is distal to the external urinary sphincter, the patient may be asymptomatic, or may complain of "vaginal voiding" wherein urine which leaks into the vagina is emptied as soon as the patient stands from a seated position. Dyspareunia and urinary tract infection (UTI) may also occur. Physical exam and cystoscopy should be performed but it may be difficult to demonstrate smaller fistulas. Voiding cystourethrogram (VCUG) is the most useful radiographic study; importantly, voiding-phase imagery must be obtained in order to diagnose urethrovaginal fistula in a patient with a competent bladder neck.

4.3 Management

Simple excision of the fistula with layered closure has a high recurrence rate, and for this reason interposition grafts are often used to reinforce the repair. A Martius labial fat pad may be tunneled beneath the labia minora and into the vaginal lumen. Use of other grafts including rectus muscle flaps, vaginal flaps, autologous fascial and synthetic grafts have all been described. Postoperatively, urinary diversion with suprapubic catheterization is more appropriate than urethral catheterization and is recommended along with a period of abstinence from sexual intercourse for 6 weeks [12]. If the initial insult causes significant damage to the bladder neck or external sphincter, fistula repair alone may fail to resolve incontinence, and in these cases a midurethral or bladder neck sling may be appropriate [12].

5. Vesicouterine fistula

5.1 Etiology

Vesicouterine fistula (VUF) is the least common form of urinary fistula, accounting for 1–4% [13, 14]. Over 75% occur after cesarean delivery [14], and there is an increase in both the absolute number of these fistulas as well as their relative proportion among urinary fistulae due to the increasing number of cesarean deliveries [15]. Other causes include prolonged labor, vaginal delivery after previous cesarean section, placenta percreta, anterior colporrhaphy, endometrial ablation, and excision or embolization of uterine fibroma [15]. Many VUF patients experience onset of symptoms with resumption of menses after childbirth, which suggests the attendant increase in circulating estrogen and resultant changes in endometrium may play a role in the pathophysiology of VUF formation [16]. In a review of 795 cases of VUF, a subset of women treated with hormonal suppression with oral contraceptives or LHRH analogs experienced resolution of their fistulae [16]. The authors speculated that hormonal suppression may be a possible treatment, but their sample size was extremely small.

5.2 Presentation and workup

Menouria is the hallmark of VUF, differentiated from hematuria as the cyclic expulsion of menstrual blood, endometrial tissue and/or blood clots with the urine. Patients may present with any combination of menouria, amenorrhea, and urinary incontinence, though not all of these features may be present simultaneously [15]. Notably, menouria and amenorrhea *without* urinary incontinence is referred to as *Youssef Syndrome*. In contrast to other urinary fistula, the absence of urinary incontinence as a reliable symptom makes the diagnosis more difficult. One proposed continence mechanism in these patients is that the uterine isthmus acts as a sphincter that restricts the passage of urine; additionally, studies have shown that the intrauterine pressure is lower than intravesical pressure for the majority of the menstrual cycle [15]. In most cases the diagnosis can be made swiftly on cystoscopy, while cystography, hysterography, and ultrasound are also often obtained.

5.3 Management

The rate of VUF resolution with bladder drainage has been estimated at 5% or less [14, 16]. Thus, the standard treatment is surgical closure using an interposition graft (omentum or peritoneum), with or without hysterectomy. While laparoscopic, transvaginal and robotic approaches have been described, the best outcomes have been reported with an open abdominal approach [15]. Given its relatively low prevalence there is a lack of studies on VUF and its impact on future fertility. Among the studies that have been published, many did not report the proportion of affected women who attempted another pregnancy. However, multiple studies have reported successful full-term pregnancies after vesicouterine fistula repair, suggesting that fertility can be recovered after treatment [15].

6. Colovesical fistula

6.1 Etiology

Colovesical fistula is an abnormal communication between the bladder and large intestine, usually sigmoid colon [17], that occurs as a sequela of inflammatory or neoplastic disease. The most common cause is diverticulitis, which accounts for 65–79% of cases [17, 18]. The relative risk of developing a colovesical fistula in the presence of diverticular disease is estimated at 2–4% [17, 18]. The next most common cause is malignancy (both colon and bladder), which accounts for 10–20% [17, 18]. Other causes include inflammatory bowel disease, radiation, appendicitis and trauma. While inflammatory bowel disease is often cited as a cause of CVF, it should be specified that it is Crohn's disease rather than ulcerative colitis that has a greater tendency to fistulize, as the entire thickness of the bowel wall is affected rather than just the mucosa. Moreover, Crohn's more often results in ileovesical fistula than CVF.

6.2 Presentation

The presenting symptoms are typically urinary, with the findings of pneumaturia (air bubbles in the urine in the absence of recent transurethral instrumentation or catheterization) and fecaluria being pathognomonic. Pneumaturia is present in 50–70% of cases [13, 17] and should be considered indicative of CVF until proven otherwise. While pneumaturia is the most common symptom, it may not be distressing to the patient and is infrequently the chief complaint. Rather, it is typically fecaluria or less-specific symptoms such as cystitis, dysuria, urinary frequency, or hematuria which drive a patient to seek medical attention. GI symptoms and leakage of urine per rectum are rare. Patients with persistent UTIs, especially those with inflammatory bowel disease, should be suspected of having a CVF and worked up appropriately [13].

6.3 Workup

6.3.1 Oral tests

There is considerable debate in the literature regarding the ideal initial workup for CVF and it should be emphasized that there is no consensus on a 'gold standard.' The classic "poppy seed test" involves oral intake of poppy seeds, which remain intact through the GI tract and may appear in the urine within 48 hours. Chen et al. [18] reported that the detection rate of the poppy seed test was 100% for surgically-confirmed fistulae, superior to CT scan. Variations of this test use activated charcoal or a variety of orally-ingested dyes. While inexpensive, these oral tests do not provide any insight into the location of the fistula or surrounding structures.

6.3.2 Contrast studies

Cystography and barium enema are likely to be abnormal in the presence of a CVF, however, the reported rates of unequivocal fistula demonstration are low at 20–34% for cystography and 30–35% for barium enema [13, 18]. Intravenous pyelogram is not as reliable for definitive diagnosis likely due to the higher intracavitary pressures generated during cystography and barium enema [13], and this test has largely fallen out of favor with the advent of newer cross-sectional imaging modalities. The *Bourne Test* refers to radiography of a centrifuged urine sample immediately after non-diagnostic barium enema and has been shown to increase the diagnostic yield of barium enema [17].

6.3.3 Cross-sectional imaging

Since the proliferation of accessible CT imaging, its diagnostic accuracy for detecting colovesical fistulae on CT has been reported at over 90%, higher than cystography or barium enema, and many have suggested this should be the initial diagnostic tool in cases of suspected colovesical fistula [17, 18]. It is recommended to perform CT after administration of oral or rectal contrast but without administration of IV contrast, as identification of enteric contrast media in the bladder provides a conclusive diagnosis. Not only is CT highly sensitive for fistula, but also it is able to rule out adjacent pathology including colonic strictures, masses, abscesses, and to localize the lesion for surgical planning. Findings on CT scan suggestive of fistula include air within the bladder lumen, as well as a thickened segment of colon in intimate contact with the bladder wall at the location of the fistula (see **Figure 1**). False positives may result after cystoscopy and in

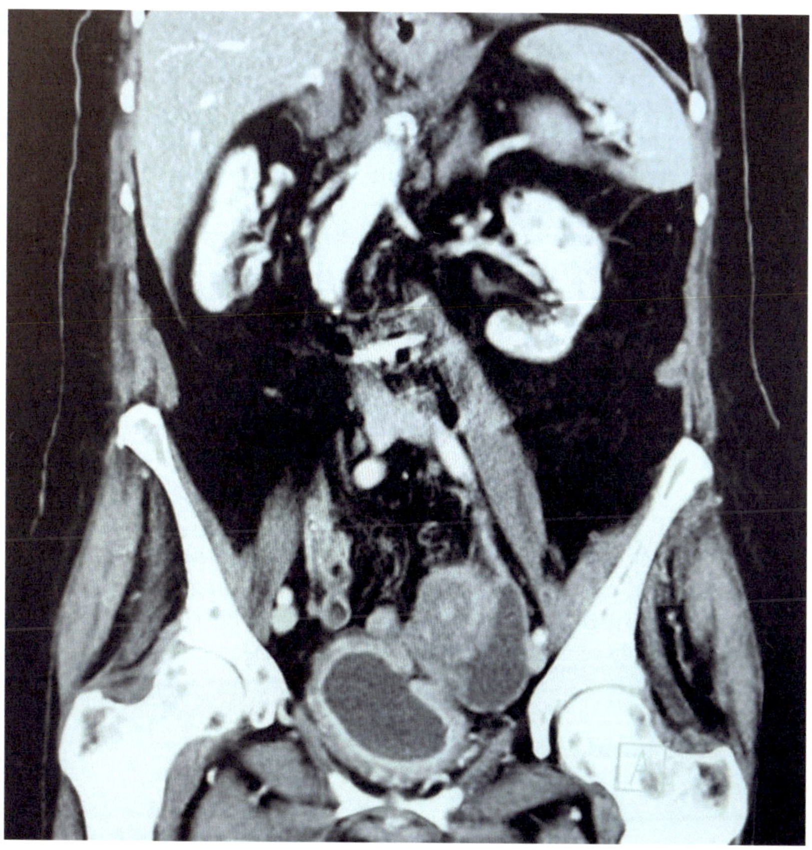

Figure 1.
CT scan showing a colovesical fistula from diverticulitis.

the presence of a gas-forming UTI or indwelling catheter; for this reason cystoscopy should be postponed until after CT. MRI has excellent sensitivity and specificity approaching 100% and is a suitable alternative to CT scan as the choice initial study [18].

6.3.4 Endoscopy

Cystoscopy is likely to be abnormal in the presence of a CVF, though findings may be nonspecific. It may reveal pieces of undigested food or feces in the bladder, which is diagnostic of CVF. Focal erythema and edema termed a *herald patch* may also indicate colovesical fistula, especially in the context of suspicious patient history and location on the bladder dome or left side nearest the sigmoid colon; however, similar lesions can be seen with intrinsic bladder pathology such as urothelial malignancy. Importantly, cystoscopy may fail to demonstrate fistula in 54–65% of cases [18] and therefore when the clinician's index of suspicion is high, the diagnosis of fistula should not be excluded on the basis of a negative cystoscopy alone. Colonoscopy has a reported detection rate of 8.5–55% for CVF [18] and is not particularly helpful in making a diagnosis. However, as 10–20% of CVF are due to bladder and bowel malignancy [17, 18], both cystoscopy and colonoscopy are considered mandatory parts of the workup regardless.

6.4 Management

Nonoperative management is an option in minimally symptomatic patients, particularly those with Crohn's. In these cases, a trial of bowel rest with parenteral nutrition, steroids or other immunomodulators along with catheter drainage is sometimes sufficient to allow for resolution of the fistula. However, due to reports of deconditioning, sepsis and increased mortality among patients with conservatively managed fistulae, this option is usually reserved for those unfit for surgery.

The recommended management in most cases involves resection of the fistula-containing segment of bowel and reconstruction of contiguous bowel anatomy. Many surgical approaches have been described including single-stage resection with primary anastomosis, as well as procedures in two stages (proximal bowel diversion and resection, followed by colostomy closure) and three stages (diversion, resection, and later closure). Advocates of the staged approach have argued that it is safest and that single-stage procedures should be reserved for patients with small fistulae of inflammatory etiology and good preoperative functional status [13]. Others have reported data suggesting that staging repairs does not reduce the risk of complications and therefore most CVF should be treated in one stage [17, 18]. An exception to either approach involves fistulae that occur after pelvic radiation, which are both more difficult to remove and more likely to recur. In these cases, a palliative bowel diversion (without resection of the fistulized bowel) may be preferable to attempting definitive repair and risking fistula recurrence and the need for repeated surgeries.

7. Ureteroarterial fistula

7.1 Etiology

Ureteroarterial fistula (UAF) is a rare condition with only a few hundred cases reported in the literature, however it is becoming more common with the increasing life expectancy of patients with pelvic malignancy, many of whom are managed with chronic ureteral stents [19, 20]. Incidence of UAF has been estimated at 2–5 patients per year [20]. Mortality has been reported to range from 7 to 23% [19]. The most common predisposing factors are pelvic surgery, radiation and ureteral stent placement which have been reported in 89–100%, 43–74% and 67–84% of UAF patients, respectively [19, 20], and all play a role in the proposed pathogenesis of UAF. Surgery and radiation causes fibrosis and fixation of the ureter to the iliac artery or aorta. Radiation causes ischemic injury to the arterial wall by damaging the vasa vasorum. Lastly, ureteral pressure necrosis results from arterial pulsation against an indwelling ureteral stent.

7.2 Presentation and workup

The main presenting symptom is hematuria, which may range from mild and intermittent to life-threatening hemorrhagic shock. Hematuria often occurs at the time of ureteral stent exchange. Flank pain has been reported in over 50% of cases [19] and may be due to obstructing ureteral blood clots. Cystoscopy may localize bleeding to one of the ureteral orifices; pulsatile bleeding on cystoscopy is indicative of UAF. Pelvic angiography is the best modality for the diagnosis and localization of UAF. Therefore, if hematuria occurs in the presence of the triad of radiation, pelvic surgery and ureteral instrumentation, the physician should consider pelvic angiography to rule out UAF [19, 20]. In a native ureter, the most common site of UAF is where the ureter crosses the bifurcation of the common iliac artery or just distal to this at proximal external iliac artery [19]. In patients with urinary diversion e.g., ileal conduit, the site of fistula may be the proximal common iliac artery or the distal abdominal aorta [19].

7.3 Management

Open surgical repair of UAF is difficult because patients have a history of pelvic surgery, radiation or both; moreover, they may be hemodynamically unstable at presentation. For these reasons, the endovascular approach has become the treatment of choice for patients with UAF regardless of age or comorbidities [19, 20]. The most common treatment is stent-graft exclusion of the fistula, with or without embolization of the affected artery and/or placement of a bypass graft. Ureteral management includes nephrostomy tube drainage, maintenance of an indwelling ureteral stent, or nephroureterectomy in the case of an atrophic or nonfunctioning ipsilateral kidney.

Lower extremity ischemia is common post-procedurally and occurs in over 50% of patients, but is somewhat less common following endovascular repair (50%) than open repair (67%) [19, 20]. Sequalae of lower extremity ischemia include skin ulcers, DVT, chronic leg pain, femoral head osteonecrosis and limb amputation. Anticoagulation has not been shown to prevent lower extremity morbidity in the setting of acute treatment of UAF. Stent-graft infection is another rare but potentially serious complication, which may be caused by either commensal skin organisms such as *Staph. Aureus* or by urinary pathogens such as *E. Coli* in the case of a UAF.

8. Conclusions

Urinary fistulae result from iatrogenic and pathologic processes leading to tissue inflammation or ischemia in or near the urinary tract. The physical and psychosocial impacts of various fistulae are often devastating for patients. Thorough history, physical exam and diagnostic testing will lead to the correct diagnosis for the patient. Conservative management may be attempted, but most patients end up undergoing definitive surgical repair. The surgical approach is tailored to each patient's anatomy and goals of care, as well as surgeon experience.

Conflicts of interest

The authors declare no conflicts of interest.

Author details

Jacob Rubin and Teresa L. Danforth*
Jacobs School of Medicine and Biomedical Sciences State University of New York at Buffalo, Buffalo, NY, United States

*Address all correspondence to: danforth@buffalo.edu

References

[1] Eilber KS, Kavaler E, Rodríguez LV, Rosenblum N, Raz S. Ten-year experience with transvaginal vesicovaginal fistula repair using tissue interposition. The Journal of Urology. 2003;**169**(3):1033-1036. DOI: 10.1097/01.ju.0000049723.57485.e7

[2] Bodner-Adler B, Hanzal E, Pablik E, Koelbl H, Bodner K. Management of vesicovaginal fistulas (VVFs) in women following benign gynaecologic surgery: A systematic review and meta-analysis. PLoS One. 2017;**12**(2):e0171554. DOI: 10.1371/journal.pone.0171554

[3] Arrowsmith S, Hamlin EC, Wall LL. Obstructed labor injury complex: Obstetric fistula formation and the multifaceted morbidity of maternal birth trauma in the developing world. Obstetrical & Gynecological Survey. 1996;**51**(9):568-574. DOI: 10.1097/00006254-199609000-00024

[4] Hilton P. Urogenital fistula in the UK: A personal case series managed over 25 years: Urogenital fistula in the UK. BJU International. 2012;**110**(1):102-110. DOI: 10.1111/j.1464-410X.2011.10630.x

[5] Hilton P, Cromwell DA. The risk of vesicovaginal and urethrovaginal fistula after hysterectomy performed in the English National Health Service--a retrospective cohort study examining patterns of care between 2000 and 2008: Risk of vesico-vaginal fistula after hysterectomy. BJOG: An International Journal of Obstetrics and Gynaecology. 2012;**119**(12):1447-1454. DOI: 10.1111/j.1471-0528.2012.03474.x

[6] Shrestha DB, Budhathoki P, Karki P, Jha P, Mainali G, Dangal G, et al. Vesico-vaginal fistula in females in 2010-2020: A systemic review and meta-analysis. Reproductive Sciences. 2022. DOI: 10.1007/s43032-021-00832-8

[7] Blaivas JG, Heritz DM, Romanzi LJ. Early versus late repair of vesicovaginal fistulas: Vaginal and abdominal approaches. The Journal of Urology. 1995;**153**(4):1110-1112; discussion 1112-3. DOI: 10.1016/s0022-5347(01)67522-0

[8] Selzman AA, Spirnak JP, Kursh ED. The changing management of ureterovaginal fistulas. The Journal of Urology. 1995;**153**(3 Pt 1):626-628. DOI: 10.1097/00005392-199503000-00020

[9] Al-Otaibi KM. Ureterovaginal fistulas: The role of endoscopy and a percutaneous approach. Urology Annals. 2012;**4**(2):102-105. DOI: 10.4103/0974-7796.95556

[10] Webster GD, Sihelnik SA, Stone AR. Urethrovaginal fistula: A review of the surgical management. The Journal of Urology. 1984;**132**(3):460-462. DOI: 10.1016/s0022-5347(17)49691-1

[11] Bahuguna G, Panwar VK, Mittal A, Talwar HS, Mandal AK, Bhadoria AS, et al. Management strategies and outcome of ureterovaginal fistulae: A systematic review and meta-analysis. Neurourology and Urodynamics. 2022;**41**(2):562-572. DOI: 10.1002/nau.24874

[12] Morse FP 3rd, Dretler SP. Diagnosis and treatment of colovesical fistula. The Journal of Urology. 1974;**111**(1):22-24. DOI: 10.1016/s0022-5347(17)59878-x

[13] Machado Junior RA, Machado Junior LC, Lourenço LLE. Vesicouterine fistula (Youssef syndrome): Case report and literature review. Revista Brasileira de Ginecologia e Obstetrícia.

2018;**40**(9):563-569. DOI: 10.1055/s-0038-1666998

[14] Jóźwik M, Jóźwik M. Spontaneous closure of vesicouterine fistula. Account for effective hormonal treatment. Urologia Internationalis. 1999;**62**(3):183-187. DOI: 10.1159/000030388

[15] Najjar SF, Jamal MK, Savas JF, Miller TA. The spectrum of colovesical fistula and diagnostic paradigm. American Journal of Surgery. 2004;**188**(5):617-621. DOI: 10.1016/j.amjsurg.2004.08.016

[16] Golabek T, Szymanska A, Szopinski T, Bukowczan J, Furmanek M, Powroznik J, et al. Enterovesical fistulae: Aetiology, imaging, and management. Gastroenterology Research and Practice. 2013;**2013**:617967. DOI: 10.1155/2013/617967

[17] Pillai AK, Anderson ME, Reddick MA, Sutphin PD, Kalva SP. Ureteroarterial fistula: diagnosis and management. AJR. American Journal of Roentgenology. 2015;**204**(5):W592-W598. DOI: 10.2214/AJR.14.13405

[18] Chen YB, Wolff BJ, Kenton KS, Mueller ER. Approach to ureterovaginal fistula: Examining 13 years of experience. Female Pelvic Medicine & Reconstructive Surgery. 2019;**25**(2):e7-e11. DOI: 10.1097/spv.0000000000000690

[19] Singh V, Mandhani PA, Mehrotra S, Sinha RJ. Laparoscopic repair of vesicouterine fistula: A brief report with review of literature. Urology Journal. 2011;**8**(2):149-152

[20] Fox JA, Krambeck A, McPhail EF, Lightner D. Ureteroarterial fistula treatment with open surgery versus endovascular management: Long-term outcomes. The Journal of Urology. 2011;**185**(3):945-950. DOI: 10.1016/j.juro.2010.10.062

Chapter 5

Endometriosis: The Enigma That It Continues to Be

Shikha Sharma and Akanksha Tripathi

Abstract

Endometriosis is an enigmatic disease and there continues to be controversy regarding its incidence, pathogenesis, history and management despite the extensive research world over. It is an estrogen-dependent, benign, inflammatory disease that affects females during their premenarcheal, reproductive, and postmenopausal hormonal stages. The prevalence in general population is not exactly known and can vary considerably depending on whether symptomatic or asymptomatic population was studied and whether clinical or surgical methods were used for diagnosis. This pathology involves presence of endometrial glands and stroma outside of the uterine cavity and the lesions additionally also, contain fibrous tissue, blood and cysts. The lesions are most commonly located in the pelvis where they can involve the ovaries, bowel as well as peritoneum and also, other sites including the diaphragm, pleural cavity, urinary system and scar sites. These lesions typically have the tendency to cause extreme anatomical distortions. Although, endometriosis is a benign pathology, ectopic endometrial tissue and resultant inflammation can cause dysmenorrhea, dyspareunia, chronic pain, and infertility. Symptoms can range from minimal to severely debilitating and may not necessarily corroborate to the degree of anatomical or architectural disarray.

Keywords: chronic pelvic pain, endometriosis, pathogenesis, medical management, surgical management

1. Introduction

Endometriosis is a chronic inflammatory condition that affects 6–10% of women of reproductive age group. It is defined by presence of endometrial like tissue outside the uterus, commonly involving the pelvic lining and structures in it including ovaries and also, sometimes sites like urinary tract, gastro-intestinal tract and even diaphgram as well as pleural cavity.

It is a benign, estrogen-dependant, inflammatory disease that has a peak prevalence in those between 25 and 35 years age group but has occasionally been reported in premenarcheal girls and postmenopausal women. Endometriosis is a common cause of subfertility and pain during periods, sexual intercourse as well as defecation.

Endometriotic lesions contain endometrial glands and stroma just as the eutopic endometrium but exist outside the uterus and often contain fibrous tissue, blood, and cysts [1]. These lesions derive pigmentation from pigmented histiocytes and hemosiderin-laden macrophages formed as a result of breakdown of red blood cells by inflammatory cells.

Endometriotic pelvic lesions can be of the following phenotypes [1, 2].

1.1 Superficial peritoneal

Although, superficial peritoneal lesions classically contain endometrial glands and stroma, either the glandular component may be absent, sparse, or transformed by hormonal, metaplastic changes and/or cellular atypia or the stromal component may be infiltrated by foamy and pigmented histiocytes, fibrosis, or other processes [3].

1.2 Ovarian endometrioma

An ovarian endometrioma occurs as a result of the ectopic endometrial tissue bleeding within the ovary and forming a hematoma surrounded by ovarian parenchyma with bilateral ovarian involvement seen in one third of cases. Unlike most haemorrhagic physiological ovarian cysts, endometriomas typically exhibit fibrotic walls with surface adhesions, contain syrup-like chocolate-colored material and lined by endometrial epithelium, stroma, and glands [2, 4]. The endometrial epithelium and stroma, more commonly, may be replaced by granulation tissue and fibrosis making histological diagnosis cumbersome.

1.3 Deep endometriosis

Deep endometriosis (DE) is when endometriotic lesion invade more than 5 mm deep to the peritoneum [5]. Most common sites are rectovaginal septum, rectum, rectosigmoid colon, bladder, ureter, and other pelvic fibromuscular structures such as the uterine ligaments and vagina.

2. Anatomy and staging

2.1 Common pelvic sites

In decreasing order of frequency are:

Ovary and ovarian fossa- 67 and 32 percent, anterior and posterior cul-de-sac, posterior broad ligaments, uterosacral ligaments- 46 percent, pouch of douglas- 30 percent, bladder- 21 percent, uterus, fallopian tubes, sigmoid colon, appendix, round ligaments, vagina, cervix, rectovaginal septum, cecum, ileum, inguinal canals, perineal scars, ureters, and umbilicus [6–9].

2.2 Extra-pelvic sites

Occasionally, endometriotic lesions may be noted in the anterior abdominal wall, usually at surgical incisional sites or umbilicus. Rarely, they have also been reported in the breast, pancreas, liver, gallbladder, kidney, urethra, extremities, vertebrae, bone, peripheral nerves, spleen, diaphragm, central nervous system, hymen and lung [10–13].

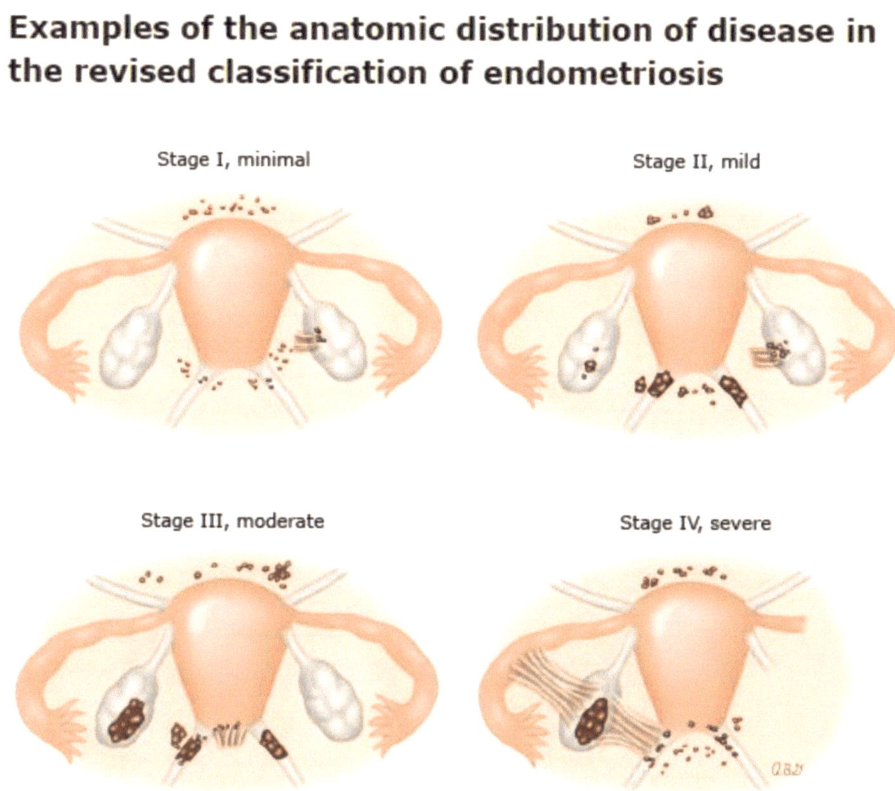

Figure 1.
ASRM classification of the disease.

2.3 Surgical staging

Endometriosis is surgically staged according to the revised American Society for Reproductive Medicine scoring system (**Figures 1** and **2**) but other staging systems also exist [14, 15]. Surgical exploration in endometriosis is indicated for

a. Evaluation of incapacitating pain or other symptoms

b. Persistent pelvic not responding to medical therapy

c. Treatment of anatomic abnormalities, such as symptomatic ovarian cysts, rectovaginal nodules, or bladder lesions.

2.3.1 Tissue biopsy versus visualization

Endometriosis is diagnosed conclusively by histologic evaluation of a visible lesion biopsied during laparoscopy [16]. Diagnosis by visual inspection alone can be limited by the stage and location of endometriosis as well as surgeon's expertise [17–19].

Table for calculating American Society for Reproductive Medicine revised classification of endometriosis

Patient's name ____________________ Date ____________

Stage I (minimal) ______ 1 to 5

Stage II (mild) ______ 6 to 15

Stage III (moderate) ______ 16 to 40

Stage IV (severe) ______ >40

Total ____________

Laparoscopy ______

Laparotomy ______

Photography ______

Recommended treatment ____________

Prognosis ____________

	Endometriosis	<1 cm	1 to 3 cm	>3 cm
Peritoneum	Superficial	1	2	4
	Deep	2	4	6
Ovary	R superficial	1	2	4
	Deep	4	16	20
	L superficial	1	2	4
	Deep	4	16	20
	Posterior cul-de-sac obliteration	Partial		Complete
		4		40
	Adhesions	<1/3 enclosure	1/3 to 2/3 enclosure	>2/3 enclosure
Ovary	R filmy	1	2	4
	Dense	4	8	16
	L filmy	1	2	4
	Dense	4	8	16
Tube	R filmy	1	2	4
	Dense	4*	8*	16
	L filmy	1	2	4
	Dense	4*	8*	16

Figure 2.
Revised ASRM classification of endometriosis.

2.3.2 Visual appearance of lesions

The gross appearance and size of visible implants can vary from raised flame-like patches, whitish opacifications, yellow-brown discolorations, translucent blebs, or reddish or reddish-blue irregularly-shaped islands or even blue- brown lesions described as "powder burns" [3]. The peritoneal surface may be scarred or puckered or have defects known as Allen-Masters syndrome, or give rise to nodules and cysts. Rarely, endometriosis may mimic a polypoidal growth giving the appearance of a malignant tumor.

2.3.3 Accuracy of biopsy

In a study of 976 women who underwent laparoscopy and biopsy for pelvic pain and/or infertility, the laparoscopic biopsy had a sensitivity of 98 percent, specificity of 79 percent, positive predictive value of 72 percent, and negative predictive value of 98 percent in diagnosing endometriosis compared with histology alone [20].

2.4 Visual lesions but negative histology

Classic endometriotic lesions at laparoscopy but with negative histology are still an indication for endometriosis treatment because negative biopsies can result from inadequate sampling and do not conclusively exclude disease [17].

2.5 Absence of visual and histologic disease

Absence of visual or histologic disease on laparoscopy excludes endometriosis with reasonable reliability [21]. Therefore, it is not advised to take random biopsies during a negative laparoscopy, although, occult microscopic sub-mesothelial implants are known to be present in normal appearing endometrium.

3. Pathogenesis

3.1 Pathogenesis and etiology

Endometriosis has a multifactorial pathogenesis, including altered immunity, imbalanced cell proliferation and apoptosis, aberrant endocrine signaling and genetic factors. Genetic studies, have identified genomic regions and mutations in cancer driver genes (PIK3CA, KRAS, ARID1A) associated with endometriosis which may partly explain the aggressive nature of deeply invasive lesions compared with superficial peritoneal lesions [22–24].

3.2 Theories of endometriosis development

Multiple existent theories for endometriosis development are,

3.2.1 Sampson's theory of retrograde menstruation

Says that endometrial cells flow backwards through the fallopian tubes and into the peritoneal cavity during menses as is evident by the increased incidence of endometriosis in girls with genital tract obstruction and increased tubal reflex. However, it is also to be noted that although, up to 90 percent of females have retrograde menstruation, most do not develop endometriosis indicating the involvement of additional factors [25–27].

3.2.2 Alternate mechanisms

Some theories suggest that the source of ectopic endometrial cells include mesothelium, stem cells, müllerian rests, bone marrow stem cells and embryonic vestiges as well as lymphatic or vascular dissemination and coelomic metaplasia [28–30].

3.2.3 Premenarcheal endometriosis

Is possibly due to the existence of müllerian embryonic rests caused by neonatal uterine bleeding resulting from maternal hormone exposure [31–33].

3.3 Mechanism of pain

The pelvic pain of endometriosis is caused by increased production of inflammatory and pain mediators as well as the associated neurological dysfunction [34]. There is a subsequent increase in nerve fibers and imbalance of sympathetic and sensory nerve fibers. The symptoms of pain may be due to the neuro-modulating effect of estrogen selectively repulsing the sympathetic axons while preserving sensory innervation, peripheral nerve sensitization, and chronic changes in the central nervous system [34–39].

3.4 Mechanism of subfertility

Endometriosis is known to affect all aspects of fertility by means of anatomic distortion from pelvic adhesions and endometriomas and/or production of substances (eg, prostanoids, cytokines, growth factors) that are "hostile" to normal ovarian function/ovulation, ova quality, sperm mobility, fertilization, and implantation.

4. Epidemiology

4.1 Prevalence

Determining the prevalence of endometriosis in general population is challenging because individuals may be asymptomatic, or have varied and non-specific presentations besides also, the fact that definitive diagnosis typically requires surgery [40].

Reported prevalence ranges for different populations are [41, 42]:

- Asymptomatic individuals- 1 to 7 percent
- Individuals undergoing hysterectomy for benign indications- 15 percent
- Adolescents with genital tract anomalies- 40 percent
- Females presenting with infertility- up to 50 percent
- Adolescent and adult females presenting for evaluation of chronic pelvic pain- up to 70 percent

4.2 Risk factors

Factors associated with an increased risk of endometriosis are positive family history, nulliparity, early menarche and late menopause, shorter menstrual cycles (defined as ≤21 days), heavy menstrual bleeding, obstruction of menstrual outflow (eg, cervical stenosis, müllerian anomalies), taller height and lower body mass index [43–51].

Factors known to reduce the risk include multiple births, extended intervals of lactation and late menarche (after age 14 years) [48, 52, 53].

4.3 Presenting symptoms

4.3.1 Common symptoms

Include chronic abdominal/pelvic pain and/or pressure, severe dysmenorrhea, dyspareunia, heavy menstrual bleeding, and infertility. The chronic pelvic pain is typically dull, throbbing, sharp, and/or burning. Additionally bowel and bladder dysfunction (eg, pain, urgency, frequency), abnormal uterine bleeding, low back pain, and chronic fatigue have also been noted [2].

4.3.2 Symptom constellation

The seven visceral symptoms associated with endometriosis include abdominal pain with no relation to menstruation, pain during urination, pain during defecation, constipation or diarrhea, irregular bleeding, nausea or vomiting, and feeling tired or lacking energy.

4.3.3 Asymptomatic individuals

Individuals with endometriosis can sometimes be asymptomatic and are often diagnosed at the time of surgery for another indication [5].

4.4 Predictive symptomatology

The type and site of endometriosis can be suggested by the leading symptom as below:

4.4.1 Dyspareunia

Dyspareunia can suggest presence of peritoneal or deep endometriosis lesions.

4.4.2 Deep dyspareunia

Deeply infiltrating endometriosis lesions can occur on the uterosacral and cardinal ligaments, pouch of Douglas, posterior vaginal fornix, and anterior rectal wall and contribute to deep sexual pain.

4.4.3 Superficial dyspareunia

Distal or superficial dyspareunia can result from lesions of the cervix, hymen, perineum, and episiotomy scars [6–10].

4.5 Urinary frequency, urgency, and/or painful micturition

Typically seen with bladder endometriosis and may worsen with menstruation. Ureteral endometriosis can be asymptomatic or associated with colicky flank pain or gross haematuria [11].

4.6 Diarrhea, constipation, and/or abdominal cramping

Typically seen with bowel endometriosis [12, 13]. Deeply infiltrating endometriosis of the posterior cul-de-sac and rectovaginal septum typically causes dyspareunia and painful defecation. Rectal bleeding may occur but is rare.

4.7 Abdominal wall pain

Cyclical or continuous abdominal wall pain is typically seen with endometriotic lesions of the abdominal wall.

4.8 Chest pain, hemoptysis, and/or pneumothorax

Individuals with thoracic endometriosis may present with chest pain, pneumothorax, haemothorax, hemoptysis, scapular or cervical pain which is mostly catamenial [15].

5. Natural history

The number of peritoneal areas affected by endometriosis appears to increase during adolescence until the early 20s [22]. Factors that cause endometriosis to either progress, regress, or remain stable are not yet fully understood.

In studies where second-look laparoscopy was performed 6 to 12 months after a diagnostic laparoscopy confirmed endometriosis, disease progressed in 29 to 45 percent of untreated women, regressed in 22 to 29 percent, and remained stable in 33 to 42 percent [23–25].

In a prospective study that followed 88 asymptomatic women with rectovaginal disease for one to nine years, fewer than 10 percent of the women had disease progression, defined as development of symptoms or increase in lesion size [26].

6. Clinical impact

6.1 Fertility

30–50% of endometriosis cases are estimated to be suffering from subfertility and 25–50% infertility cases are estimated to have endometriosis [2, 3]. It is hypothesized that minimal/mild endometriosis incites an inflammatory response involving overproduction of prostaglandins, metalloproteinases, cytokines, and chemokines as well as macrophages and natural killer cells. This inflammatory process is known to impair ovarian, peritoneal, tubal, and endometrial function leading, subsequently, to defective folliculogenesis, fertilization as well as implantation.

Several studies have reported that the increased numbers of macrophages and cytokines in the peritoneal fluid of women with endometriosis inhibits both sperm and fallopian tube ciliary function in vitro. In addition, the eutopic endometrium also, may not function with endometriosis as a consequence of which, implantation is impaired [5, 6].

In a study involving monkeys with experimentally induced endometriosis, the pregnancy rate was approximately 40 percent in normal controls, but only 12 percent in animals with advanced endometriosis and 0 percent if ovarian adhesions were present [8].

Advanced endometriosis is also reported to adversely impact the quality of ovarian follicles leading to abnormal folliculogenesis and reduced fertilization potential of oocytes [9, 10].

6.2 Pregnancy

Often causes decidualization of the endometriotic lesions leading to reduction or disappearance in the endometriosis related pain. However, complications have been reported in even the decidualized spots such as intestinal perforation, hemoperitoneum, uroperitoneum, acute appendicitis, and ruptured or infected ovarian endometrioma. Since these events are a rarity, it is not warranted to do any additional monitoring for pregnant individuals. Possible explanations could be traction by the growing uterus on adhesions, increased friability of inflamed tissues, and alteration of vessel walls by decidualized lesions [54–63].

6.3 Obstetric outcomes

Evidence suggests an increased risk of preterm birth, preeclampsia, and cesarean birth when compared with individuals without endometriosis (adjusted odds ratio [aOR] 1.67, 95% CI 1.37–2.05; aOR 1.37, 95% CI 1.06–1.77; aOR 1.83, 95% CI 1.60–2.09, respectively). A meta-analysis of 33 studies on over 3 million pregnant women reported that, amongst those who conceived spontaneously, endometriosis was associated with placenta previa, cesarean delivery, preterm birth, and low birth weight. However, additional surveillance for pregnant individuals with endometriosis is not advised [63–73].

6.4 Postmenopause

Although, endometriosis may remain active after menopause with approximately 2 percent of all initial surgical diagnosis seen amongst postmenopausal women, severe symptomatic endometriosis is not usually seen in that age group. Symptomatic postmenopausal endometriosis has been seen amongst those both on and off hormone therapy. This could possibly be due to extra-ovarian estrogen production (eg, skin, fat tissue) and or lesion-specific production [74, 75].

6.5 Ovarian cancer risk

6.5.1 Summary

It has been reported that ovarian endometriosis, whether superficial lesions or endometriomas or both, could have an increased risk of ovarian cancer, particularly endometrioid and clear cell types. Some ovarian lesions harbor genetic changes that are common to ovarian cancer such as mutations in ARID1A, PTEN, HNF1B and k-ras. While there is a raised relative risk, the absolute risk of ovarian cancer remains low. The risk of malignant transformation in endometriosis is estimated at 1 percent and 1–2.5 percent for premenopausal and post-menopausal women respectively. In a meta-analysis of 13 case-control studies including nearly 8000 women with epithelial ovarian cancer (EOC), those with a history of endometriosis had three times the risk of clear cell EOC and two times the risk of endometrioid and low-grade serous EOC but no change in risk of high-grade serous or mucinous EOC. Ovarian cancer developing in endometriotic lesions is seen in younger women (49 versus 59 years old) and with a better medial overall survival (196 versus 34 months) than patients without endometriosis. Activation of oncogenic *KRAS* and *PI3K* pathways and inactivation of tumor suppressor genes *PTEN* and *ARID1A* have been suggested as mechanisms for the transformation of endometriosis, particularly ovarian endometriomas, to malignancy [76–81].

6.5.2 Ovarian cancer surveillance not advised

Endometriosis is not considered a premalignant lesion and surveillance for ovarian cancer is not recommended. Also, there is no evidence to suggest prophylactic removal of endometriosis lesions. However, oral contraceptive pills lower the risk of ovarian cancer in all users.

6.5.3 Cardiovascular disease

Two studies reported an increased risk of cardiovascular disease, myocardial infarction, composite cardiovascular disease, and all-cause mortality in those harboring endometriosis [82, 83]. However, more data is needed to validate this fact.

7. Medical treatment options

Include nonsteroidal analgesics, hormonal contraceptives, gonadotropin-releasing hormone (GnRH) analogues, and aromatase inhibitors (AI). Medical interventions do not improve fertility, diminish endometriomas, or treat those features of deep endometriosis that require targeted definitive management strategies.

7.1 Nonsteroidal anti-inflammatory drugs

Nonsteroidal anti-inflammatory drugs (NSAIDs), e.g. Propionic acids such as ibuprofen, naproxen or Fenamates such as mefenamic acid, acetic acids such as Indomethacin, diclofenac, etodolac, Oxicams such as piroxicam are considered the first-line treatment for pelvic pain, including endometriosis-related pain as they are affordable, readily available, carry an acceptable side-effect profile, and have consistently demonstrated effective reduction of primary dysmenorrhea. For women attempting conception, we avoid selective Cyclo-oxygenase-2 (COX-2) inhibitors (celecoxib, refecoxib, and valdecoxib) as some studies indicate these drugs can prevent or delay ovulation [2].

7.2 Estrogen-progestin contraceptives

Combined contraceptives are an excellent option as they can be used long-term, are affordable, well-tolerated, easy to use, as well as provide additional benefits of contraception and ovarian and endometrial cancer risk reduction [84]. Estrogen-progestin contraceptives suppress ovarian function causing decidualization and subsequent atrophy of endometrial tissue.

Although, both cyclic and continuous-dose hormonal regimens can be used, two systematic reviews reported that continuous combined oral contraceptive

(COC) regimens were more effective at pain reduction than cyclical COC regimens [12, 13, 84].

7.3 Progestins

For women with contra-indications to estrogens, progestin-only therapy is a valid treatment option. The available options are:

- **Norethindrone acetate**- Commonly used as 5 mg daily, but can be increased to 15 mg daily depending on side effects including breakthrough bleeding [15].
- **Medroxy-progesterone acetate**- Depot MPA can be given as 150 mg intramuscular injection every three months or as a 104 mg subcutaneous injection every three months [23].
- **Dienogest**- Prescribed commonly as a 2 mg oral daily pill or combined with estradiol valerate as a multiphasic oral contraceptive pill [24]. Also, available in 1 mg and 4 mg doses.

Progestins act by causing decidualization and atrophy of endometrial tissue and additionally, also, cause suppression of matrix metalloproteinases and inhibition of angiogenesis.

Progestin-only treatment avoids the estrogen-related thromboembolic risk as seen with estrogen-containing contraceptives and also, avoids bone loss seen with GnRH analogues. When compared with danazol, progestins are better tolerated with less impact on lipids and no androgenic side effects. However, progestins have earned a little disrepute as they are known to cause irregular uterine bleeding/spotting, amenorrhea, weight gain, mood changes, and bone loss specifically with long-term use of depot MPA as well as 4 mg dose of dienogest. Additionally, long-term use of norethindrone acetate causes significant reduction in high-density lipoprotein with subsequent increase in low-density lipoprotein, cholesterol and triglycerides necessitating monitoring of lipid levels in patients on long-term therapy [1, 26, 27].

7.4 Levonorgestrel intrauterine device (LNG- IUS)

A review of three trials concluded that the recurrence of dysmenorrhoea in women with surgically confirmed endometriosis reduced with the postoperative use of LNG-IUS significantly [28, 29]. However, irregular menstrual bleeding and amenorrhea were common side effects, with no effect on bone mineral density or lipid levels [30, 31].

7.5 Gonadotropin-releasing hormone agonists/antagonists

7.5.1 Agonist analogues

GnRH agonists include nafarelin, leuprolide, buserelin, goserelin, and triptorelin. A meta-analysis reported that GnRH analogs were more effective than placebo and as effective as other medical therapies (danazol, levonorgestrel, COC) for relieving pain [85]. A trial comparing GnRH agonists treatment, laparoscopy, and combined medical/surgical treatment concluded an equal cure rate of ≥50 percent for all three groups.

Commonly used regimens of GnRH agonists include leuprolide acetate 3.75 mg intramuscular monthly, leuprolide acetate 11.25 mg intramuscular every three months, and intranasal nafarelin acetate 200 mcg given twice daily with the choice being driven by availability and cost [86].

GnRH agonists bind to receptors in the pituitary gland leading to down-regulation of the pituitary-ovarian axis which causes hypoestrogenism, amenorrhea and progressive endometrial atrophy. The hypoestrogenic state is what causes the pain relief but on the flip side also, causes adverse effects, including hot flushes, vaginal dryness, decreased libido, mood swings, headache, and loss of bone density. Add-back therapy, typically with oral norethindrone acetate or a combination of estrogen and progestin (COC pill) diminishes the hypoestrogenic side effects to a great extent [1, 87–89].

7.5.2 Antagonist

GnRH antagonists suppress the pituitary gonadotropin hormone production causing a hypoestrogenic stat. As antagonists do not cause an initial surge in LH and FSH, unlike agonists, they do not require 7 to 14 days for GnRH suppression. The antagonists elagolix and relugolix are available in oral form [34]. The GnRH antagonists induce a dose-dependent hypoestrogenic state to inhibit endometriotic cell proliferation. Symptom relief and adverse events such as vasomotor phenomena, vaginal atrophy, and bone loss are all dose-dependent.

7.5.3 Elagolix

Available in doses of 150 mg to be taken once daily for up to 24 months (low dose) or 200 mg twice daily for up to six months (high dose). Although, the lower effective dose is preferred, higher dose works better in cases of severe deep dyspareunia. The medication is started within seven days of menses. It is to be noted that low-dose elagolix seemed less effective as compared to GnRH agonist.

Significant reduction in dysmenorrhoea was reported by 44, 74 and 21 percent of women in the high-dose, low dose and placebo groups respectively while non-menstrual pelvic pain was decreased in 50, 56, and 36 percent of women in the low-dose, high-dose and placebo groups respectively.

7.5.4 Adverse events and side effects

Hepatic dysfunction is an absolute contra-indication while the reported side effects include hot flushes or night sweats. Hot flushes were the most common adverse event.

Mean deviation from baseline in lumbar spine bone density, at 6 month interval, were approximately −0.5, −2.6, and + 0.5 percent in the low-dose, high-dose, and placebo groups, respectively.

7.5.5 Relugolix

Can be used as combination therapy (relugolix 40 mg, estradiol 1 mg, and norethisterone acetate 0.5 mg) or as monotherapy (40 mg dose for 12 weeks) followed by combination therapy for 12 weeks.

7.5.6 Relugolix phase-three trial data

Two phase-three, multicenter trials, performed across six continents, compared placebo with relugolix combination therapy (relugolix 40 mg, estradiol 1 mg, and norethisterone acetate 0.5 mg) or delayed relugolix combination therapy (relugolix monotherapy, 40 mg dose for 12 weeks, followed by combination therapy for 12 weeks) in individuals with moderate-to-severe endometriosis-related pain. Across the two trials, compared with the placebo group, more patients receiving relugolix reported reductions in dysmenorrhea (75 percent versus 27 to 30 percent) and less non-menstrual pelvic pain (58 to 66 percent versus 40 to 43 percent) at six months of follow-up. The most common adverse events were headache, nasopharyngitis, and hot flushes. Both treatment groups do cause some loss of bone mineral density but it is insignificant. Data comparing relugolix with established treatments for endometriosis-related pain are lacking.

7.6 Newer GnRH antagonists

7.6.1 Linzagolix

It is a new oral, non-peptide GnRH antagonist under investigation currently. The pharmacokinetics, pharmacodynamic impact on estradiol and safety of this drug have already been studied in a previous trial on healthy female volunteers [90]. In order to study the effect of Linzagolix on endometriosis associated pain (EAP), a multinational, parallel group, randomized, placebo-controlled, double blind was carried out on women aged 18–45 years with surgically proven endometriosis. The interventions examined were 50, 75, 100 or 200 mg linzagolix administered daily for a period of 24 weeks.

Estimated percentages of women with a $\geq$ 30% reduction in overall pelvic pain (OPP) at week 12 were 34.5%, 49.4%, 61.5%, 56.4%, and 56.3% in the placebo, 50 mg, 75 mg, 100 mg, and 200 mg groups, respectively. The differences, as compared with placebo, were significant in the 75 mg ($p = .003$), 100 mg ($P = .039$), and 200 mg ($P = .034$) dose groups but not in the 50 mg dose group ($P = .155$). There was a significant reduction in dyschezia scores in the 75 mg ($P = .002$), 100 mg ($P = .004$), and 200 mg ($P = .022$) dose groups at 12 weeks. Dyspareunia scores at week 12 decreased significantly compared with placebo in the 200 mg dose group (P=. 023) [91].

7.6.2 ASPI707

ASPI707 is an oral GnRH antagonist currently being investigated for endometriosis-related pain, which it is known to ameliorate by supressing the pituitary-gonadal axis. Phase I studies on healthy young women suggest that it effectively suppresses E2 levels, thereby, reducing endometriosis- related pain. The phase II TERRA study evaluated the efficacy, safety and dose-response relationship of this drug in subjects with endometriosis- associated pelvic pain. This trial was a multinational, multicentre, double-blind, randomized, parallel group, placebo- controlled study conducted in Europe and Japan wherein the eligible subjects were randomized in a ratio of 1:1:1:1:1:1 to placebo, one of four doses of ASPI707 (3,5,10 or 15 mg) or leuprolide acetate (subcutaneous injection 3.75 mg/month) observed over a 24 week treatment period. Compared with placebo, statistically significant differences in the absolute change from baseline in mean NRS were observed for OPP (10 mg, $P = 0.011$) and dysmenorrhea (3 mg, $P = 0.003$; 5 mg, $P < 0.001$; 10 mg, $P < 0.001$; 15 mg, $P < 0.001$).

Although, Leuprorelin and all ASP1707 dose groups showed statistically significant reductions in bone marrow density (BMD) from baseline at 24 weeks compared with baseline, the percentage reduction across all ASP1707 dose groups was significantly lower than leuprorelin for total hip ($P \leq 0.017$) and spine ($P \leq 0.001$). The effect of ASP1707 on the BMD of the spine was lower for the 10-mg dose group compared with the 15-mg dose group [92].

7.6.3 Danazol

Danazol is a 17 alpha-ethinyl testosterone derivative that inhibits steroidogenesis, luteinizing hormone surge and increases free testosterone levels. It primarily acts by inhibiting pituitary gonadotropin secretion, directly inhibiting ovarian enzymes secretion and thereby, curbing growth of endometriotic implants.

It is highly effective at treating endometriosis-related pain, with various studies validating that six months of danazol treatment was more effective than placebo in relieving endometriosis-related pain and that the effect persisted even six months after discontinuation of therapy. However, the unwelcome side effects, including acne, muscle cramps, oedema, weight gain of up to 5 percent of body weight, spotting, hirsutism, and voice deepening have limited the use of danazol. Danazol is typically given orally in divided doses ranging from 400 to 800 mg daily, generally for six months. One systematic review reported symptom relief with vaginal danazol, particularly for cases with rectovaginal endometriosis [38–42].

7.6.4 Aromatase inhibitors

Endometriosis treatment is an off-label use of these medications. It is usually reserved for women with severe, refractory, endometriosis-related pain. AIs are often used in combination with progestins. Typical treatments include oral anastrozole 1 mg once daily or oral letrozole 2.5 mg once daily, usually in combination with progestins. These act by regulating local estrogen formation within the endometriotic lesions, in addition to inhibiting estrogen production in the ovary, brain, and peripheral tissues (eg, adipose tissue) [40].

A systematic review concluded that AI treatment, with or without progestin, COC, or GnRH analogue, significantly reduced pain compared with GnRH analogue alone. An RCT comparing six months of post-operative goserelin plus anastrozole with goserelin alone in severe endometriosis cases reported longer time to symptom recurrence with combination therapy but equal quality-of-life and bone mass outcomes between the groups [44].

Side effects of prolonged AIs use include bone loss and ovarian follicular cyst development. It is therefore, prescribed in combination with a GnRH analogue or an oral estrogen-progestin contraceptive or oral norethindrone acetate 5 mg alone to suppress follicular development.

8. Surgical treatment

Surgical treatment is reserved for cases wherein the medical treatment has either failed or the symptoms have re-surfaced. Surgical resection, besides pain relief, provides tissue for histological diagnosis [1]. However, endometriosis surgery involves substantial risk of injury to the bowel and bladder, possible reduction of ovarian reserve in case of endometriomas and post-operative adhesion formation, in addition to the common surgical risks.

8.1 Surgical resection

Surgical resection may either be conservative involving targeted ablation or resection of visible lesions, or definitive involving hysterectomy with or without oophorectomy in addition to resection of endometriosis or radical that involves removal of all visible implants.

8.2 Nerve transection

Includes laparoscopic uterosacral nerve ablation (LUNA) and presacral neurectomy (PSN). The available data suggests that LUNA is ineffective. Although, PSN may be effective for midline pain, but the success of this procedure depends largely on the experience of the surgeon, and it's long-term efficacy remains questionable [1].

9. Special populations

9.1 Infertility

The treatment of infertility associated with endometriosis involves a combination of surgery and assisted reproduction technology. Hormonal therapies, alone, are not offered as they prevent pregnancy and do not improve fertility.

9.2 Endometrioma

Endometriomas are managed to relieve pain, exclude malignancy, improve subfertility, and preserve ovarian function. Symptomatic or expanding endometriomas are removed laparoscopically, which is known to diminish ovarian reserve [45, 46]. It is a wise clinical practice to not operate on asymptomatic and small endometriomas $\leq$5 cm.

9.3 Deep endometriosis

The management of rectovaginal or bowel endometriosis is guided by the symptomatology, location of lesions, and patient expectation of the treatment. For women with bothersome urinary or bowel symptoms such as urinary urgency or frequency, dyspareunia, dysmenorrhea, or dyschezia, hormonal suppression is appropriate [49]. Surgery is better preserved for women with ureteral or bowel obstruction or women whose symptoms do not improve with medical management [3].

9.4 Lesions of nonreproductive organs

Endometriosis has been reported at distant locations like upper abdomen, diaphragm, abdominal wall (umbilicus, surgical scar), perineum (episiotomy or obstetric scar), and thorax [3]. Complications such as pneumothorax are treated as indicated.

Typically, ovarian suppression with gonadotropin-releasing hormone (GnRH) analogues is the first line of treatment since, they effectively suppress ovarian hormone production and also, inhibit the growth of endometrial tissue [50–58].

10. Complementary therapies

10.1 Acupuncture

Acupuncture may serve as a complement or alternative to the treatment modalities mentioned. Acupuncture is believed to alleviate pain by means of various physiological and psychological processes, such as activation of endogenous descending pain inhibitory systems, deactivation of brain areas that transmit pain-related signals, interaction between nociceptive impulses and somato-visceral reflexes, and the expectation of symptom relief [93, 94].

Various studies have demonstrated that acupuncture can reduce the level of serum CA-125, thereby, relieving the pelvic pain associated with endometriosis [95–98]. In a trial on 67 participants, auricular acupuncture was found to be significantly more effective than Chinese herbal medicine for treating dysmenorrhea in women with endometriosis [59, 60]. Current literature consistently finds that acupuncture yields better reductions in pain and serum CA-125 levels than do control treatments, regardless of the control intervention used. To confirm this finding, additional studies with proper controls, blinding methods, and adequate sample sizes are needed.

10.2 Diet

Endometriosis, as we all know, is a multifactorial pathology with a concomitant inflammatory pattern. Although, high level evidence and well-designed randomized studies on the impact of modifiable risk factors on endometriosis are lacking, certain studies have suggested that fish oil may have a positive effect on reducing endometriosis related pain. This may be due the anti-inflammatory prostaglandins PGE3 and PGE3 $\propto$ which are derivatives of omega-3 fatty acids. Similarly, antioxidants vitamins C, D and E are also, seen to alleviate endometriosis symptoms of pain [99].

One study reported that higher intake of green vegetables and fruit reduced the risk of developing endometriosis while higher intake of beef or other red meat increased the endometriosis risk [61]. Several studies have addressed the correlation of diet and dysmenorrhea, but not exclusively in patients with endometriosis. There is also evidence that the adaptation of individualized dietary changes yields statistically significant improvements in endometriosis-related symptoms [100]. Thus, there may be great benefit to including a dietician in the management of these patients, especially at younger ages and in early stages.

11. Conclusion

Endometriosis is a pathology that is known to affect women of all age groups including pre-menarcheal as well as post-menopausal. It can have an array of presentations from asymptomatic to subfertility to incapacitating symptoms. The management of endometriosis includes mainly medical treatment while surgical management is preserved for those who fail to respond to medical management or require correction of tubo-ovarian disarray.

12. Summary and recommendations

1. Pelvic endometriosis can be categorized as superficial peritoneal, ovarian and deep depending upon the size and depth of the lesions.

2. Globally, ten percent of reproductive age group women have endometriosis but the prevalence may vary as per clinical population.

3. The pathogenesis appears to be multifactorial including ectopic endometrial tissue, altered immunity, imbalanced cell proliferation and apoptosis, aberrant endocrine signaling, and genetic factors.

4. Endometriosis is staged at the time of surgery and most commonly the American Society for Reproductive Medicine scoring is used.

5. It is associated with reduced fertility, poor obstetrical outcomes, epithelial ovarian cancer especially clear cell carcinoma and endometroid ovarian cancer and also, atherosclerosis.

6. In general, the management of endometriosis depends upon patients' symptomatology and their expectation of the same.

7. Surgical treatment is only preserved for cases of failed medical management and those requiring repair of tubo-ovarian architecture.

Author details

Shikha Sharma[1*] and Akanksha Tripathi[2]

1 Western Command Hospital, Chandimandir, India

2 Ananta Institute of Medical Sciences and Research Centre, Udaipur, Rajasthan, India

*Address all correspondence to: shikhalhmc@gmail.com

References

[1] International working group of AAGL, ESGE, ESHRE and WES, Tomassetti C, Johnson NP, et al. An international terminology for endometriosis, 2021. Journal of Minimally Invasive Gynecology. 2021;**28**:1849

[2] Vercellini P, Viganò P, Somigliana E, Fedele L. Endometriosis: Pathogenesis and treatment. Nature Reviews. Endocrinology. 2014;**10**:261

[3] Clement PB. The pathology of endometriosis: A survey of the many faces of a common disease emphasizing diagnostic pitfalls and unusual and newly appreciated aspects. Advances in Anatomic Pathology. 2007;**14**:241

[4] Muzii L, Bianchi A, Bellati F, et al. Histologic analysis of endometriomas: What the surgeon needs to know. Fertility and Sterility. 2007;**87**:362

[5] De Cicco C, Corona R, Schonman R, et al. Bowel resection for deep endometriosis: A systematic review. BJOG. 2011;**118**:285

[6] Audebert A, Petousis S, Margioula-Siarkou C, et al. Anatomic distribution of endometriosis: A reappraisal based on series of 1101 patients. European Journal of Obstetrics, Gynecology, and Reproductive Biology. 2018;**230**:36

[7] Victory R, Diamond MP, Johns DA. Villar's nodule: A case report and systematic literature review of endometriosis externa of the umbilicus. Journal of Minimally Invasive Gynecology. 2007;**14**:23

[8] Jaime TJ, Jaime TJ, Ormiga P, et al. Umbilical endometriosis: Report of a case and its dermoscopic features. Anais Brasileiros de Dermatologia. 2013;**88**:121

[9] Lee HJ, Park YM, Jee BC, et al. Various anatomic locations of surgically proven endometriosis: A single-center experience. Obstetrics & Gynecology Science. 2015;**58**:53

[10] Minaglia S, Mishell DR Jr, Ballard CA. Incisional endometriomas after cesarean section: A case series. The Journal of Reproductive Medicine. 2007;**52**:630

[11] Horton JD, Dezee KJ, Ahnfeldt EP, Wagner M. Abdominal wall endometriosis: A surgeon's perspective and review of 445 cases. American Journal of Surgery. 2008;**196**:207

[12] Morales Martínez C, Tejuca SS. Abdominal wall endometriosis. American Journal of Obstetrics and Gynecology. 2017;**217**:701

[13] Canlorbe G, Laas E, Cortez A, Daraï E. Spontaneous hymeneal endometriosis: A rare cause of dyspareunia. BMJ Case Reports. 2014:1-2. DOI: 10.1136/bcr-2013-202299

[14] Abrao MS, Andres MP, Miller CE, et al. AAGL 2021 endometriosis classification: An anatomy-based surgical complexity score. Journal of Minimally Invasive Gynecology. 2021;**28**:1941

[15] Keckstein J, Saridogan E, Ulrich UA, et al. The #Enzian classification: A comprehensive non-invasive and surgical description system for endometriosis. Acta Obstetricia et Gynecologica Scandinavica. 2021;**100**:1165

[16] Duffy JM, Arambage K, Correa FJ, et al. Laparoscopic surgery for endometriosis. Cochrane Database of Systematic Reviews. 2014;**4**:CD011031

[17] Dunselman GAJ et al. ESHRE guideline: Management of women with endometriosis. **29**(3):400-412. Available from: https://www.eshre.eu/Guidelines-and-Legal/Guidelines/Endometriosis-guideline

[18] Stegmann BJ, Sinaii N, Liu S, et al. Using location, color, size, and depth to characterize and identify endometriosis lesions in a cohort of 133 women. Fertility and Sterility. 2008; **89**:1632

[19] Fernando S, Soh PQ, Cooper M, et al. Reliability of visual diagnosis of endometriosis. Journal of Minimally Invasive Gynecology. 2013;**20**:783

[20] Almeida Filho DP, Oliveira LJ, Amaral VF. Accuracy of laparoscopy for assessing patients with endometriosis. São Paulo Medical Journal. 2008;**126**:305

[21] Wykes CB, Clark TJ, Khan KS. Accuracy of laparoscopy in the diagnosis of endometriosis: A systematic quantitative review. BJOG. 2004;**111**: 1204

[22] Rahmioglu N, Nyholt DR, Morris AP, et al. Genetic variants underlying risk of endometriosis: Insights from meta-analysis of eight genome-wide association and replication datasets. Human Reproduction Update. 2014;**20**:702

[23] Anglesio MS, Papadopoulos N, Ayhan A, et al. Cancer-associated mutations in endometriosis without cancer. The New England Journal of Medicine. 2017;**376**:1835

[24] Bulun SE, Wan Y, Matei D. Epithelial mutations in endometriosis: Link to ovarian cancer. Endocrinology. 2019;**160**:626

[25] Sampson JA. Peritoneal endometriosis due to the menstrual dissemination of endometrial tissue into the peritoneal cavity. American Journal of Obstetrics and Gynecology. 1927;**14**:422

[26] Olive DL, Henderson DY. Endometriosis and mullerian anomalies. Obstetrics and Gynecology. 1987;**69**:412

[27] Dovey S, Sanfilippo J. Endometriosis and the adolescent. Clinical Obstetrics and Gynecology. 2010;**53**:420

[28] Burney RO, Giudice LC. Pathogenesis and pathophysiology of endometriosis. Fertility and Sterility. 2012;**98**:511

[29] Du H, Taylor HS. Contribution of bone marrow-derived stem cells to endometrium and endometriosis. Stem Cells. 2007;**25**:2082

[30] Gruenwald P. Origin of endometriosis from the mesenchyme of the celomic walls. American Journal of Obstetrics and Gynecology. 1942;**44**:470

[31] Marsh EE, Laufer MR. Endometriosis in premenarcheal girls who do not have an associated obstructive anomaly. Fertility and Sterility. 2005;**83**:758

[32] Oliveira FR, Dela Cruz C, Del Puerto HL, et al. Stem cells: Are they the answer to the puzzling etiology of endometriosis? Histology and Histopathology. 2012;**27**:23

[33] Figueira PG, Abrão MS, Krikun G, Taylor HS. Stem cells in endometrium and their role in the pathogenesis of endometriosis. Annals of the New York Academy of Sciences. 2011;**1221**:10

[34] McKinnon BD, Bertschi D, Bersinger NA, Mueller MD.

Inflammation and nerve fiber interaction in endometriotic pain. Trends in Endocrinology and Metabolism. 2015; **26**:1

[35] Arnold J, Barcena de Arellano ML, Rüster C, et al. Imbalance between sympathetic and sensory innervation in peritoneal endometriosis. Brain, Behavior, and Immunity. 2012;**26**:132

[36] Arnold J, Vercellino GF, Chiantera V, et al. Neuroimmunomodulatory alterations in non-lesional peritoneum close to peritoneal endometriosis. Neuroimmunomodulation. 2013;**20**:9

[37] Liang Y, Yao S. Potential role of estrogen in maintaining the imbalanced sympathetic and sensory innervation in endometriosis. Molecular and Cellular Endocrinology. 2016;**424**:42

[38] Wheeler MA, Heffner DL, Kim S, et al. TNF-α/TNFR1 signaling is required for the development and function of primary nociceptors. Neuron. 2014; **82**:587

[39] As-Sanie S, Harris RE, Napadow V, et al. Changes in regional gray matter volume in women with chronic pelvic pain: A voxel-based morphometry study. Pain. 2012;**153**:1006

[40] Hickey M, Ballard K, Farquhar C. Endometriosis. BMJ. 2014;**348**:g1752

[41] old Shafrir AL, Farland LV, Shah DK, et al. Risk for and consequences of endometriosis: A critical epidemiologic review. Best Practice & Research. Clinical Obstetrics & Gynaecology. 2018;**51**:1

[42] Soliman AM, Surrey E, Bonafede M, et al. Real-world evaluation of direct and indirect economic burden among endometriosis patients in the United States. Advances in Therapy. 2018;**35**: 408

[43] Falcone T, Flyckt R. Clinical Management of Endometriosis. Obstetrics and Gynecology. 2018;**131**:557

[44] Ballard KD, Seaman HE, de Vries CS, Wright JT. Can symptomatology help in the diagnosis of endometriosis? Findings from a national case-control study–part 1. BJOG. 2008; **115**:1382

[45] Hediger ML, Hartnett HJ, Louis GM. Association of endometriosis with body size and figure. Fertility and Sterility. 2005;**84**:1366

[46] Sinaii N, Plumb K, Cotton L, et al. Differences in characteristics among 1,000 women with endometriosis based on extent of disease. Fertility and Sterility. 2008;**89**:538

[47] Giudice LC. Clinical practice. Endometriosis. The New England Journal of Medicine. 2010;**362**:2389

[48] Treloar SA, Bell TA, Nagle CM, et al. Early menstrual characteristics associated with subsequent diagnosis of endometriosis. American Journal of Obstetrics and Gynecology. 2010;**202**: 534.e1

[49] Nnoaham KE, Webster P, Kumbang J, et al. Is early age at menarche a risk factor for endometriosis? A systematic review and meta-analysis of case-control studies. Fertility and Sterility. 2012;**98**:702

[50] Benagiano G, Brosens I. In utero exposure and endometriosis. The Journal of Maternal-Fetal & Neonatal Medicine. 2014;**27**:303

[51] Farland LV, Missmer SA, Bijon A, et al. Associations among body size

across the life course, adult height and endometriosis. Human Reproduction. 2017;**32**:1732

[52] Parazzini F, Cipriani S, Bianchi S, et al. Risk factors for deep endometriosis: A comparison with pelvic and ovarian endometriosis. Fertility and Sterility. 2008;**90**:174

[53] Farland LV, Eliassen AH, Tamimi RM, et al. History of breast feeding and risk of incident endometriosis: Prospective cohort study. BMJ. 2017;**358**:j3778

[54] Pisanu A, Deplano D, Angioni S, et al. Rectal perforation from endometriosis in pregnancy: Case report and literature review. World Journal of Gastroenterology. 2010;**16**:648

[55] Brosens IA, Fusi L, Brosens JJ. Endometriosis is a risk factor for spontaneous hemoperitoneum during pregnancy. Fertility and Sterility. 2009; **92**:1243

[56] Brosens IA, Lier MC, Mijatovic V, et al. Severe spontaneous hemoperitoneum in pregnancy may be linked to in vitro fertilization in patients with endometriosis: A systematic review. Fertility and Sterility. 2016;**106**:692

[57] Lier M, Malik RF, van Waesberghe J, et al. Spontaneous haemoperitoneum in pregnancy and endometriosis: A case series. BJOG. 2017;**124**:306

[58] Lier MCI, Malik RF, Ket JCF, et al. Spontaneous hemoperitoneum in pregnancy (SHiP) and endometriosis - a systematic review of the recent literature. European Journal of Obstetrics, Gynecology, and Reproductive Biology. 2017;**219**:57

[59] Chiodo I, Somigliana E, Dousset B, Chapron C. Urohemoperitoneum during pregnancy with consequent fetal death in a patient with deep endometriosis. Journal of Minimally Invasive Gynecology. 2008;**15**:202

[60] Leone Roberti Maggiore U, Remorgida V, Sala P, et al. Spontaneous Uroperitoneum and preterm delivery in a patient with bladder endometriosis. Journal of Minimally Invasive Gynecology. 2015;**22**:923

[61] Faucheron JL, Pasquier D, Voirin D. Endometriosis of the vermiform appendix as an exceptional cause of acute perforated appendicitis during pregnancy. Colorectal Disease. 2008;**10**: 518

[62] Murphy SJ, Kaur A, Wullschleger ME. Endometrial decidualization: A rare cause of acute appendicitis during pregnancy. Journal of Surgery Case Reports. 2016; **2016**(4):1-3

[63] Leone Roberti Maggiore U, Ferrero S, Mangili G, et al. A systematic review on endometriosis during pregnancy: Diagnosis, misdiagnosis, complications and outcomes. Human Reproduction Update. 2016;**22**:70

[64] Glavind MT, Forman A, Arendt LH, et al. Endometriosis and pregnancy complications: A Danish cohort study. Fertility and Sterility. 2017;**107**:160

[65] Saraswat L, Ayansina DT, Cooper KG, et al. Pregnancy outcomes in women with endometriosis: A national record linkage study. BJOG. 2017;**124**:444

[66] Stephansson O, Kieler H, Granath F, Falconer H. Endometriosis, assisted reproduction technology, and risk of adverse pregnancy outcome. Human Reproduction. 2009;**24**:2341

[67] Fernando S, Breheny S, Jaques AM, et al. Preterm birth, ovarian endometriomata, and assisted reproduction technologies. Fertility and Sterility. 2009;**91**:325

[68] Santulli P, Marcellin L, Menard S, et al. Increased rate of spontaneous miscarriages in endometriosis-affected women. Human Reproduction. 2016;**31**: 1014

[69] Exacoustos C, Lauriola I, Lazzeri L, et al. Complications during pregnancy and delivery in women with untreated rectovaginal deep infiltrating endometriosis. Fertility and Sterility. 2016;**106**:1129

[70] Bruun MR, Arendt LH, Forman A, Ramlau-Hansen CH. Endometriosis and adenomyosis are associated with increased risk of preterm delivery and a small-for-gestational-age child: A systematic review and meta-analysis. Acta Obstetricia et Gynecologica Scandinavica. 2018;**97**:1073

[71] Farland LV, Stern JE, Liu CL, et al. Pregnancy outcomes among women with endometriosis and fibroids: Registry linkage study in Massachusetts. American Journal of Obstetrics and Gynecology. 2022;**226**:829.e1

[72] Marcellin L, Goffinet F, Azria E, et al. Association between endometriosis phenotype and preterm birth in France. JAMA Network Open. 2022;**5**:e2147788

[73] Lalani S, Choudhry AJ, Firth B, et al. Endometriosis and adverse maternal, fetal and neonatal outcomes, a systematic review and meta-analysis. Human Reproduction. 2018;**33**:1854

[74] ESHRE Guideline Endometriosis. European Society of Human Reproduction and Embryology. 2022. https://www.eshre.eu/Guidelines-and-Legal/Guidelines/Endometriosis-guideline [Accessed: February 18, 2022]

[75] Attar E, Bulun SE. Aromatase and other steroidogenic genes in endometriosis: Translational aspects. Human Reproduction Update. 2006; **12**:49

[76] Saavalainen L, Lassus H, But A, et al. Risk of gynecologic cancer according to the type of endometriosis. Obstetrics and Gynecology. 2018;**131**:1095

[77] Oxholm D, Knudsen UB, Kryger-Baggesen N, Ravn P. Postmenopausal endometriosis. Acta Obstetricia et Gynecologica Scandinavica. 2007;**86**: 1158

[78] Van Gorp T, Amant F, Neven P, et al. Endometriosis and the development of malignant tumours of the pelvis. A review of literature. Best Practice & Research. Clinical Obstetrics & Gynaecology. 2004;**18**:349

[79] Pearce CL, Templeman C, Rossing MA, et al. Association between endometriosis and risk of histological subtypes of ovarian cancer: A pooled analysis of case-control studies. The Lancet Oncology. 2012;**13**:385

[80] Orezzoli JP, Russell AH, Oliva E, et al. Prognostic implication of endometriosis in clear cell carcinoma of the ovary. Gynecologic Oncology. 2008; **110**:336

[81] Grandi G, Toss A, Cortesi L, et al. The association between Endometriomas and ovarian cancer: Preventive effect of inhibiting ovulation and menstruation during reproductive life. BioMed Research International. 2015;**2015**:751571

[82] Mu F, Rich-Edwards J, Rimm EB, et al. Endometriosis and risk of coronary heart disease. Circulation.

Cardiovascular Quality and Outcomes. 2016;**9**:257

[83] Okoth K, Wang J, Zemedikun D, et al. Risk of cardiovascular outcomes among women with endometriosis in the United Kingdom: A retrospective matched cohort study. BJOG. 2021;**128**: 1598

[84] Gustofson RL, Kim N, Liu S, Stratton P. Endometriosis and the appendix: A case series and comprehensive review of the literature. Fertility and Sterility. 2006;**86**:298

[85] Brosens I, Puttemans P, Benagiano G. Endometriosis: A life cycle approach? American Journal of Obstetrics and Gynecology. 2013;**209**: 307

[86] Wang G, Tokushige N, Markham R, Fraser IS. Rich innervation of deep infiltrating endometriosis. Human Reproduction. 2009;**24**:827

[87] Tran LV, Tokushige N, Berbic M, et al. Macrophages and nerve fibres in peritoneal endometriosis. Human Reproduction. 2009;**24**:835

[88] Matsuzaki S, Darcha C. Involvement of the Wnt/β-catenin signaling pathway in the cellular and molecular mechanisms of fibrosis in endometriosis. PLoS One. 2013;**8**:e76808

[89] Adamson GD, Kennedy S, Hummelshoj L. Creating solutions in endometriosis: Global collaboration through the world endometriosis research foundation. Journal of Endometriosis. 2010;**2**:3-6

[90] Pohl O, Marchand L, Fawkes N, Gotteland JP, Loumaye E. Gonadotropinreleasing hormone receptor antagonist mono- and combination therapy with estradiol/ norethindrone acetate add-back: Pharmacodynamics and safety of OBE2109. The Journal of Clinical Endocrinology and Metabolism. 2018; **103**:497-504

[91] Donnez J, Taylor HS, Taylor RN, et al. Treatment of endometriosis-associated pain with linzagolix, an oral gonadotropin-releasing hormone-antagonist: A randomized clinical trial. Fertility and Sterility. 2020;**114**(1):44-55

[92] D'Hooghe T, Fukaya T, Osuga Y, et al. Efficacy and safety of ASP1707 for endometriosis-associated pelvic pain: The phase II randomized controlled TERRA study. Human Reproduction. 2019;**34**(5):813-823

[93] Hui KK, Liu J, Marina O, Napadow V, Haselgrove C, Kwong KK, et al. The integrated response of the human cerebro-cerebellar and limbic systems to acupuncture stimulation at ST 36 as evidenced by fMRI. NeuroImage. 2005;**27**(3):479-496. DOI: 10.1016/ j.neuroimage.2005.04.037

[94] Chen S, Wang S, Rong P, Wang J, Qiao L, Feng X, et al. Acupuncture for visceral pain: Neural substrates and potential mechanisms. Evidence-based Complementary and Alternative Medicine. 2014;**2014**:609594. DOI: 10.1155/2014/609594

[95] Sun YZ, Chen HL. Controlled study on Shu-Mu point combination for treatment of endometriosis. Zhongguo Zhen Jiu. 2006;**26**(12):863-865 Chinese

[96] Xiang DF, Sun QZ, Liang XF. Effect of abdominal acupuncture on pain of pelvic cavity in patients with endometriosis. Zhongguo Zhen Jiu. 2011; **31**(2):113-116 Chinese

[97] Xu Y, Zhao W, Li T, Zhao Y, Bu H, Song S. Effects of acupuncture for the

treatment of endometriosis-related pain: A systematic review and meta-analysis. PLoS ONE. 2017;**12**(10):e0186616. DOI: 10.1371/journal.pone.0186616

[98] Zhang XX, Li W. Efficacy on endometriosis treated with electroacupuncture. Zhongguo Zhen Jiu. 2015;**35**(4):323-326 Chinese

[99] Habib N, Buzzaccarini G, Centini G, Moawad GN, Ceccaldi PF, Gitas G, et al. Impact of lifestyle and diet on endometriosis: A freshlook to a busy corner. Menopause Review/Przegląd Menopauzalny. 2022;**21**(2):124-132

[100] Karlsson JV, Patel H, Premberg A. Experiences of health after dietary changes in endometriosis: A qualitative review study. BMJ Open. 2020;**10**: e032321

Chapter 6

Pelvic Floor Awareness and Adult Re-Toilet Training in the Treatment of Lower Urinary Tract Dysfunction

Nicole Fleischmann

Abstract

Toileting habits are established early in life, during the process of toilet training. Children who overuse their guarding reflex can develop voiding dysfunction and incontinence, which they typically outgrow. However, unhealthy behaviors, such as enhanced holding and straining to void, are common in the adult general population, suggesting the persistence of old habits that have the potential to create symptoms throughout a woman's life. Women with poor toileting habits frequently ignore bodily sensations, and use voiding postponement and abdominal straining to eliminate, without relaxing the pelvic floor. Continuous pressure on the pelvic floor tissues during elimination can create structural changes in the pelvic anatomy, particularly after childbirth. Educating women about pelvic floor function, bladder retraining, and full pelvic floor relaxation during voiding and defecating are the goals of treatment. In this chapter, we review a simple and strategic process in which the provider teaches relaxation of the puborectalis muscle, with the aid of diaphragmatic breathing, so that coordinated elimination can develop. Although pelvic floor awareness and education may not benefit all cases, women should be warned that chronic pelvic muscle clenching and straining are potentially destructive to the pelvic floor and sphincter muscle function.

Keywords: pelvic floor dysfunction, lower urinary tract dysfunction, urinary incontinence, external urethral sphincter, toilet posture, behavioral therapy

1. Introduction

Female pelvic floor disorders are pervasive and costly problems leading not only to diminished quality of life, but severe health issues including overactive bladder, urinary incontinence, painful bladder syndrome, pelvic pain, dyspareunia, recurrent urinary tract infections, and pelvic organ prolapse. While we have come to understand the causes of these common conditions to be multifactorial, with contributions from childbearing, aging, genetics and lifestyle, little emphasis has been placed on the role of toileting habits. In the last two decades, there has been an increased understanding that women who exhibit unhealthy habits such as delayed voiding, straining to void or positional voiding have higher rates of lower urinary tract dysfunction (LUTD) including overactive bladder symptoms, incomplete bladder emptying, recurrent urinary tract infections and urinary leakage [1–4]. A cross sectional study of the US population showed 76% of women admit to delayed voiding and 62% strain to void, leading the authors to conclude that "this work is important for the development of future public health interventions to encourage healthy toileting behaviors and equitable access to toileting environments that support bladder health" [5].

Many investigators have proposed that adult toileting behaviors are persistent learned childhood behaviors. Children who delay voiding also have a propensity for bowel dysfunction, which motivated the International Continence Society (ICS) to suggest the umbrella term, "bladder and bowel dysfunction (BBD)", to describe these related conditions [6]. Many children with BBD display abnormal or unhealthy toileting habits which are highly sustained into adulthood [7–10]. There is a significant association of lower urinary tract symptoms (LUTs) and a history of unhealthy toileting habits such as infrequent urination, postponement of urination, and urinary urgency in childhood [11]. Childhood dysfunctional voiding may predict adult bladder control problems particularly in women with urge incontinence [12]. Pelvic floor disorders and LUTs are approximately doubled in nulliparous women with a history of childhood bedwetting [13]. Nocturnal enuresis in childhood is predictive of nocturia and frequency in adult women [14]. The long-term follow-up of children with nocturnal enuresis in a large cohort of 516 patients and showed that, despite the resolution of bedwetting, a third of patients go on to experience nocturia and a fourth of them still report LUTs [15]. A recent study demonstrated that, according to ICS criteria, 40% of females who had extensive urotherapy for voiding dysfunction in childhood had persistent voiding dysfunction as an adult, 56% still experienced daytime incontinence and 28% had recurrent UTI's. The results of these studies suggest that the resolution of childhood LUTs and incontinence does not necessarily indicate the resolution of the underlying pathological condition [16].

Other factors which contribute to toileting habits are occupational and lifestyle constraints [17]. Poor access to toilet facilities undoubtedly leads to prolonged holding which is beyond one's control [18]. There is also the problem of lack of public education about what is considered "healthy toileting". The majority of women are uninformed about how often they should be using the bathroom and how to position/posture themselves on the toilet to promote effective emptying [19]. Due to concerns about hygiene, many women hover or squat over public toilets, a position known to promote dysfunctional habits such as straining to urinate or "power peeing" [20]. Furthermore, women are surprisingly unaware of their own toileting behaviors and are known to be poor historians when asked specific questions about their voiding habits [21]. They also demonstrate a poor knowledge about the pelvic floor and related dysfunctional conditions, expressing a strong desire for more information [22, 23].

Toileting habits and lower urinary tract anatomy illustrate the dynamic relationship between function and structure, which interplay over the course of a lifetime. One of the key principles of functional anatomy is that this relationship is bidirectional—a behavior can change the structure of a body part, which in turn affects the function of that part [24]. In the case of toileting habits, there may be anatomic changes in the lower urinary tract caused by chronic straining to eliminate which change the structure of the system to accommodate the behavior. As we will discuss later in this chapter, stress urinary incontinence may start with straining behaviors causing anatomic changes in the pelvic floor which facilitate voiding by weakening the outlet. Additionally, the abdominal strain necessary to overcome the obstruction of a non-relaxed outlet can create an anterior vaginal wall defect, causing further outlet obstruction from pelvic organ prolapse [25]. Because of the variability of outcomes, establishing the relationship between toileting behaviors and pelvic floor disorders is challenging. The salient questions to answer are (1) will correction of toileting habits through extensive education make a difference in the prevalence of pelvic floor disorders? and (2) what is the most effective way to provide that educational support.? If we are able to look at the vast amount of evidence already present in the literature, spanning multiple fields including adult and pediatric urology, gastroenterology, neurology and behavioral psychology, we can develop a greater understanding of the scope of the problem and further insight into answering these questions. In this chapter we will review the process of acquiring bladder control and how dysfunctional habits develop. We will then discuss a theory for how toileting habits may be the primary underlying cause for adult pelvic floor disorders including urinary incontinence and pelvic organ prolapse. Lastly, we discuss re-toilet training techniques with an emphasis on pelvic floor awareness.

2. The development of lower urinary tract function and dysfunction (LUTD)

The principal role of the lower urinary tract is to allow for low pressure storage of urine and its complete expulsion from the urinary bladder without having to generate high pressures that could be transmitted to the kidneys. Although the anatomic structures, including the urinary bladder and the bladder outlet, are rich in peripheral nerve pathways, the coordinated activity of these structures is mediated by a complex neural control system located in the central nervous system. The initiation of micturition is achieved through voluntary relaxation of the striated urogenital sphincter complex, followed almost immediately by a detrusor contraction and complete bladder emptying [26]. Coordinated voiding is not present at birth so that neonates and infants have uncontrolled urination and do not empty their bladders completely [27]. Social continence is achieved through physiological maturation and parental guidance, which, in the United States, occurs between 3 and 5 years of age, when toilet training is usually complete [28]. At this milestone, the child is presumably (1) aware of inner body sensations that the bladder is full (2) able to initiate toilet finding behaviors in a timely fashion (3) able to contract the pelvic floor muscles until the toilet is found (4) able to empty the bladder in a coordinated fashion.

The process by which controlled urination occurs is not fully understood but is thought to be mediated by use and disuse of the "guarding reflex". Children learn to purposefully inhibit reflex voiding and to consciously initiate desired voiding using the external sphincter as an "on and off switch" [29]. Voluntary contraction of the external sphincter, suppresses detrusor activity by stimulation of the sympathetic (holding) and inhibition of parasympathetic (releasing) responses of the lower urinary tract [26]. As children enter structured environments, such as school, they must navigate time schedules and circumstances which can distract them from noticing internal bodily sensations until they come on suddenly, necessitating extreme holding maneuvers such as leg crossing and "Vincent's curtesy" to suppress the urge [30]. When holding behaviors are consistently prioritized at the expense of appropriate outlet relaxation during voiding, dysfunctional patterns emerge which create bladder outlet obstruction (BOO) [31]. While the majority of children acclimate with maturity, LUTD is the basis for 40% of all pediatric urology visits [32]. A study of 2856 school-age children, 5–20% had daytime/nighttime wetting with the female to male ratio of 3:1 [33]. In severe cases, such as Hinman's syndrome, obstructive voiding patterns lead to incomplete emptying, chronically elevated bladder pressures causing reflux to the kidneys and the potential for renal loss [34]. Fowlers syndrome is another condition seen predominantly in young females with a poorly relaxing sphincter and reflexively acontractile detrusor muscle, resulting in urine retention. A key feature is decreased bladder sensation, with failure to process internal signals of fullness causing large capacity bladders [35].

Enhanced holding behaviors appear to be influenced by psychological processes and emotional stress. In a cohort of 62 women with Fowler's syndrome, 24% had medically unexplained somatic symptoms that included loss of consciousness, limb weakness, sensory disturbance, and memory impairment. Furthermore, almost a third (31%) had psychological comorbidities such as anxiety/depression or obsessive-compulsive symptoms [36]. The latter conditions were found in 29% of children with urinary incontinence vs. 6% in healthy controls [37]. Delayed voiding is associated with an increased rate of psychosocial difficulties as well as stressful childhood experiences. These include divorce, bullying, physical and sexual abuse, adoption, moving multiple times, and death of a family member [38]. In a large epidemiologic study of 8213 children aged 7.5–9 years, children with daytime wetting had significantly increased rates of psychological problems especially separation anxiety, attention deficit, oppositional behavior, and conduct problems [39]. Studies have shown that relapse of nocturnal enuresis in children can be induced by stressful life changes such as separation and divorce [40]. Similar patterns are noted in adults with histories of childhood voiding dysfunction indicating there is an ability to regress to immature voiding patterns during periods of increased psychosocial stress [41]. One study showed that nearly 50% of patients with OAB have anxiety and or depression, with 25% having severe symptoms [42]. Proposed mechanisms for anxiety causing bladder and bowel dysfunction are based on animal studies demonstrating stress related activation of the hypothalamic-pituitary axis, which releases cortisol and adrenaline in a sympathetic response, overly enhancing outlet tone and worsening LUTs [43]. Urinary incontinence can itself be the source of anxiety and depression, suggesting that the relationship is cyclical [44, 45]. Chronic sympathetic response initiates a vicious cycle in which an individual, distressed by their incontinence, has increased psychological disturbance which perpetuates the incontinence. Regardless of the underlying mechanism, undue psychological stress during development is a proven barrier to functional voiding.

3. Diagnosing LUTD in adult females

The obstructive mechanism of LUTD in adults is similar to that in children. Incomplete relaxation of the outlet during the voiding phase can result in typical overactive bladder (OAB) symptoms such as frequency, urgency, nocturia and urge incontinence. Women who have managed to effectively suppress bladder contraction with a strong muscular outlet can develop detrusor underactivity (DU), characterized by a normal or increased capacity with "contraction of reduced strength and/or duration, resulting in prolonged bladder emptying and/or a failure to achieve complete bladder emptying within a normal time span". According to the ICS, underactive bladder (UAB) describes a slow urinary stream, hesitancy and straining to void with or without a feeling of incomplete bladder emptying, sometimes with storage symptoms [46]. Both OAB and UAB can present with storage and emptying symptoms, requiring pressure flow studies to make the distinction, although there are no standardized parameters to confirm either diagnosis.

UAB is a common finding in women with voiding dysfunction but there is little information on the true prevalence, although it is known to be more common in the aging population [47]. The diagnosis of UAB is made when LUTS such as frequency, urgency and incomplete bladder emptying are paired with decreased voiding pressures on urodynamic testing. Uroflowmetry may show a low flow with staccato interrupted pattern and elevated post void residual or a completely normal curve with complete emptying [48]. Nitti et al. demonstrated that it is typical for women to void using less than 10 cm of water pressure to completely empty with a good urinary flow rate using subtle maneuvers such as abdominal straining [49]. Straining is a common voiding pattern in women—with and without LUTs—seen in as much as 75% of the general population of women [48]. Strain patterns are easy to overlook on pressure flow studies especially when there is a normal bell-shaped flow curve. Someone who pushes or bears down to urinate can generate substantial flow rates to rapidly empty her bladder. Despite its efficiency, investigators have warned that abdominal straining during the entire micturition process can have deleterious outcomes [50]. In one study, half of women without subjective complaints of voiding difficulty (poor stream, straining to void, hesitancy or incomplete emptying) voided with abdominal straining as a compensation for a compressed or constricted urethral outlet [3]. Investigators have found that strain and postural changes are risk factors for recurrent UTIs, lower urinary tract symptoms, pelvic organ prolapse, incontinence and bladder outlet obstruction [51].

Straining to void is seen frequently in women with stress incontinence. In a large study of 1788 patients, those with DU were more likely to report stress urinary incontinence, reduced stream, straining to void, enuresis, absent or reduced sensation, previous urinary retention and use of antidepressants [52]. When urodynamic studies in patients with and without SUI were compared, significantly lower voiding pressures were noted in women evaluated for SUI than in the normal controls [53]. Similarly, eighty-four percent of women with genuine SUI voided without a detrusor contraction compared to none of the patents without SUI [54]. In women with SUI, dysfunctional patterns such as straining and staccato voiding may be hidden by an incompetent sphincter which could obscure the diagnosis of bladder outlet obstruction (BOO). Strain voiding rarely presents with bothersome voiding symptoms and women may only seek consultation for the accompanying SUI [55]. The underlying voiding patterns are important to consider when correcting SUI because straining to void is associated with higher rates of urinary retention and lower success rates after anti-incontinence surgery. Strain voiders are more likely to develop urgency after pubovaginal sling [56]. Women who go into retention after PVS may fail the surgery by loosening the sling through bearing down behaviors [57].

Straining maneuvers are also a risk factor for pelvic organ prolapse (POP) [58]. Forty-one percent of women with POP have DU, voiding symptoms and incomplete emptying, with larger defects having a greater likelihood of underlying DU and voiding dysfunction [59]. Chronic constipation is highly associated with all degrees of POP. In one study, after controlling for important covariates, straining at stool remained associated with anterior vaginal wall and perineal descent [60]. When researchers compared urodynamic patterns of patients with SUI, with and without anterovaginal wall prolapse, women with prolapse had higher urethral resistance during the voiding phase, even with cystocele reduction. The authors concluded that patients with prolapse have DU with elevated BOO that is not related to a structural abnormality [61]. The generation of POP is a good demonstration of the complex interplay between structure and function. Women who have obstructed elimination patterns develop weakness in their vaginal walls through straining which then worsen the obstruction by further compressing the opening. Without knowing the "pre-prolapse" voiding pattern, clinicians may assume the obstruction is caused by the prolapse, and not the reverse.

4. Understanding pelvic floor function and dysfunction (PFD)

The pelvic floor muscles work in tandem with the anal and urethral sphincters to control urination and defecation. The levator ani muscles—the pubococcygeus (PCM) and iliococcygeus (ICM) muscle—comprise the deep floor. The superficial muscles, which are caudal to the deep floor, are the urethrovaginal sphincter (UVS), compressor urethrae (CU) muscle, the ischiocavernosus, and bulbospongiosus muscles complex [62]. The puborectalis (PR), often considered part of the superficial floor, is a muscular sling with two well-developed parallel fascicles that insert anteriorly in the pubis, and fuse together behind the rectum at the top of the anal canal. The traction of the PR maintains anorectal angulation and places pressure on the opening of the anus, urethra and vagina to constrict these openings against the pubic bone [63]. The deep muscles of the pelvic floor serve to support the pelvic organs, working synergistically with the thoracic diaphragm and abdominal wall to maintain low intrabdominal pressure (IAP). The superficial muscles, including the UVS and CU components of the external urinary sphincter (EUS) and the PR muscle are concentrically placed and work in tandem to compress the three openings of the pelvic floor—the urethra, the vagina and the rectum—during elevated IAP. This constriction is often involuntary, as during positional changes, creating a high pressure zone of the vagina [64]. As in the case of the lower urinary tract, higher cortical functions are needed to balance voluntary control with the involuntary contraction necessary for activities and positional changes [65].

Pelvic floor dysfunction is a generalized term which describes a group of muscular conditions characterized by either hypotonic (weak) or hypertonic (overly tense) muscles. The classic teaching is that women with hypotonic muscles have SUI and POP whereas women with hypertonic muscles have irritative symptoms and pelvic pain syndromes, although this may be an oversimplification. In 2021, the ICS published an updated list of terms which are used in the neuromyofascial assessment and diagnosis of pelvic floor muscle disorders to aid teaching and standardization of terminology in this field. Hypertonic pelvic floor disorders are defined by increased "resting tension clinically determined by resistance to passive movement" [66]. Elevated pelvic floor tone and myofascial levator tenderness is seen in 87% of patients with chronic pelvic pain and IC/PBS [67]. The development of myalgia maybe related to the degree of spasm with higher tone causing higher pain levels. Chronic elevated resting tension is often cited as the basis for the majority of pelvic floor pain and dyspareunia, causing local impairment of blood circulation and the exchange of oxygen and metabolites, leading to dysmorphic changes in muscle structure and neuropathic pain [68].

While many clinicians are familiar with pelvic floor "tension", chronic pelvic muscle clenching (CPMC) is an unconscious, habitual overuse of the guarding reflex. The ICS has acknowledged the need for a new term, "involuntary contraction" which is defined as "pelvic floor contraction which occurs reflexively or automatically, without volition or conscious control" [66]. Involuntary movements in the pelvic floor help support continence and elimination, however a significant portion of activity may also be utilized to manage emotional states. Psychological stress leads to chronically high tone in the pelvic floor [69, 70]. Women are typically unaware of increased pelvic tone nor do they associate the state of their pelvic muscles with their symptoms. Some women continuously tighten their muscles for fear of losing urine or stool. The cycle of increased outlet tone causing worsening BOO and LUTs, leading to reflexive increase in voluntary pelvic muscle clenching to prevent leakage is a common occurrence. Clenching is also used as a protective response to pain in the visceral area, or if pain is anticipated causing further tightening and further discomfort [68].

For reasons not clearly understood, sensorimotor control of the pelvic floor muscles is underdeveloped in many women. Only a minority of women are able to correctly engage their muscles despite confidence that they are exercising correctly [71]. Multiple reasons for the diminished proprioception seen in women with and without pelvic floor disorders have been proposed, including an inability to visualize or "map" the pelvic floor muscles, lack of information about pelvic floor anatomy, and conflicting messages given to women from a very young age. There is also considerable cultural, racial and generational differences in women's comfort levels with discussion on the subject of urogenital functions [23]. Unawareness of the pelvic floor muscles is closely associated with pelvic floor disorders. Fifty-three percent of women with SUI are unable to perform coordinated pelvic floor movements with simple verbal instructions [72]. When women with and without functional constipation were compared, those without obstruction had a better ability to identify and relax their pelvic floor muscles [73]. Regardless of their condition, clinicians routinely counsel women to perform Kegel's or pelvic floor tightening exercises without emphasizing pelvic relaxation [74]. Women with high tone pelvic floors can experience increased pelvic pain and worsening organ dysfunction if they engage in unsupervised strengthening exercises [75]. Rigorous exercise programs such as cross fit, which emphasize core strengthening, can have the adverse effect of enhancing pelvic spasm by incorporating the pelvic floor [76].

The challenge for the clinician is to be able to diagnose high tone PFD and when there is extreme variability in its presentation. Female pelvic muscle strength is difficult to assess, particularly when there is an overall weakness in the vaginal walls, as in the case with POP. Some women with pelvic muscle tears, weakened deep levator muscles and widened genital hiatus can still have high tone/outlet pressure even though there is little resistance to the examiners finger on palpation. Some women who appear to have weakened strength have so much muscle tension at "rest" that they are unable to further contract to execute an effective squeeze around the examiner's finger. While a physical examination of the myofascial structures can show elevated resting tone of the puborectalis muscle, as determined by digital palpation, there is poor reliability [77]. Numerous scoring systems have been developed which rely on subjective quantification of tone and function and have not been validated [78]. Other assessment techniques, including transperineal ultrasound, manual muscle testing, and squeeze pressure measurements, have similarly produced unreliable results [79]. Levator tenderness is a sufficient but not a necessary criteria for high tone PFD as many cases do not result in pelvic floor myalgia or allodynia [80]. Due to the likelihood of discoordination of the pelvic floor muscles in LUTD, high tone PFD and CPMC are the diagnoses to be excluded in all patients with LUTs and incontinence.

5. A functional theory of SUI

LUTD leads to strain assisted voiding patterns which have the potential to cause distortion in the vaginal wall including bladder neck hypermobility and pelvic organ prolapse, particularly after childbirth. Even small amounts of continuous pressure on the weakened pelvic support tissues may prevent damaged fascia, ligaments and muscles from regaining their nulliparous structure [81, 82]. Delancey proposed that stress urinary incontinence develops from damage to the support of the urethra which requires a strong backboard to maintain function [83]. As described by Petros and Ulmsten in their integral theory, the mechanism is made possible by several anatomical structures. Most importantly, the pubourethral ligaments (PUL) attach the pubic bone to the midurethra and the external urethral ligaments (EUL) affix the muscular components of the distal urinary sphincter—the CU and UVS—to the undersurface of the symphysis pubis. The vaginal hammock is a fascial condensation extending between these two ligamentous structures. The integrity of the PUL allows the distal muscles of the EUS to approximate with the urethral crest (UC), the cavernosal type tissue of the distal urethrovaginal septum which is analogous to the verumontanum in males. When the CU and UVS contract, they exert a downward force on the lumen of the urethra, compressing this tissue and creating a strong seal for continence. The configuration is important in that it implies that the closure of the urethra by the EUS is done slit-wise by compressing the ventral part against the dorsal part instead of closing in a purely circular way. The authors referred to this mechanism as the 'first closure mechanism' (**Figure 1**) [84].

Women with chronic DU who use straining against a narrow outlet to void are at risk for developing sphincter damage and SUI after a vaginal delivery. Loss of urethral support through vaginal distortion compromises the alignment of the CU and UVS over the UC which prevents the occlusive seal needed to prevent leakage when the IAP rises. A closer look at this mechanism is demonstrated in **Figure 2**. The EUS muscles are stabilized by the EUL which remains relatively unharmed by childbirth. In contrast, the UC is part of the distal anterior vaginal, normally stabilized by intact PULs which are primarily injured in vaginal delivery. In women with SUI, during cough or valsalva, there is rotational movement of the external meatus around the intact EUL causing the UC to become displaced (**Figure 2(2)** and **(3)**). When the EUS muscles clamp down to make contact with the UC, the target is missed causing an insufficient

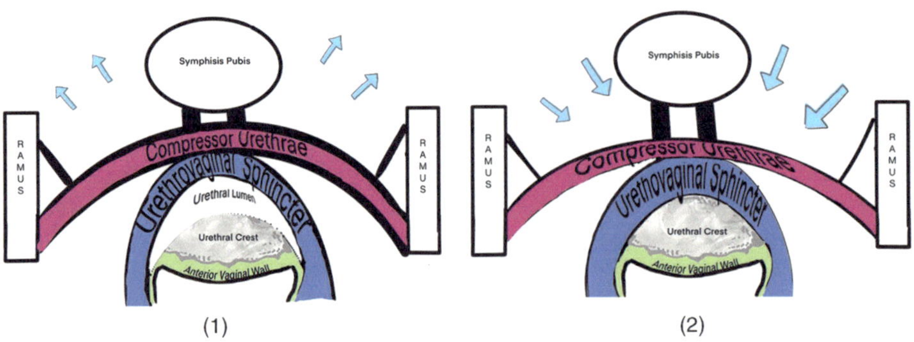

Figure 1.
External urinary sphincter closure mechanism. (1) Open EUS during pelvic floor relaxation and functional urination and (2) downward force of EUS muscles on urethral crest creates continence seal.

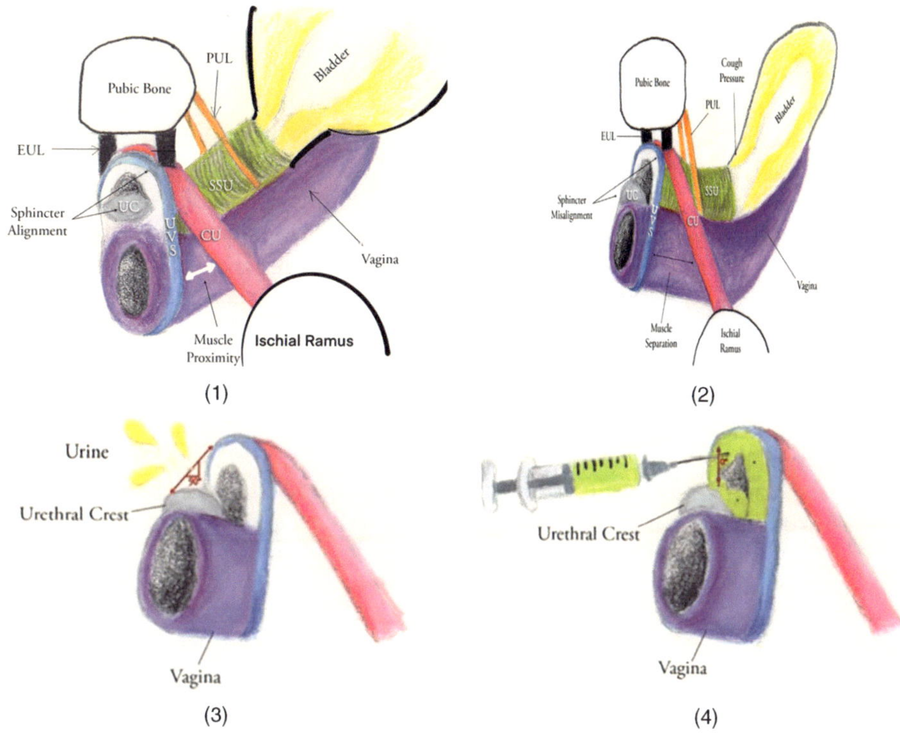

Figure 2.
External sphincter complex misalignment. (1) Pre-cough position—compressor urethrae (CU) and urethrovaginal sphincter are aligned with urethral crest (UC), (2) sphincter misalignment with cough. EUL = external urethral ligament, PUL = pubourethral ligament, SSU = striated sphincter urethrae, (3) 90 degree muscle misalignment during cough causing leakage of urine and (4) reconstruction of EUS with polyacrylamide hydrogel.

occlusive seal. The amount of movement required to produce leakage depends on multiple factors such as striated muscle atrophy as seen with aging, estrogen deficiency, genetic tissue quality, urethral scarring, fibrosis and denervation of the EUS muscles. Purportedly, the MUS procedure works to realign the sphincter component by preventing the rotational motion that destabilizes the EUS complex.

In the current model, poor apical support would also compromise EUS alignment, and therefore, the function of the first closure mechanism. In the integral theory, the "second closure mechanism" relies on both functioning PULs as well as apical support of the vaginal wall. Poor proximal urethra support, as exists when there is significant apical prolapse, prevents the necessary backward motion of the levator plate which then fails to stretch the proximal urethra at the zone of critical elasticity (ZCE), impairing bladder neck closure. This plate stretching may also be important for EUS alignment. The authors stated that "an intact first closure mechanism is an important prerequisite for the proper functioning of the second closure mechanism" [84]. Elevating the apical anterior vaginal wall in patients who have significant bladder neck hypermobility, would potentially reduce the separation of the UVS and CU, and pull the urethral crest back into approximation with the EUS muscles. In patients with prolapse and SUI undergoing mid-urethral sling, patients with tighter USL repairs had better continence outcomes than those with loose USL repairs [85]. The idea of apical support facilitating EUS closure seems counterintuitive when one considers the similarly appearing but entirely distinct mechanism of unmasking occult stress incontinence in patients who have apical prolapse repairs. In these scenarios, prolapse correction unobstructs the bladder neck, increasing the flow to and unveiling a weakly functioning EUS. Therefore, EUS closure is also dependent on substantial tissues for creating a seal. Realignment of the EUS through vaginal suspension alone will not work to cure incontinence in the setting of striated muscle damage.

As we further delineate the pathophysiology of SUI, potential new treatment modalities emerge. For example, urethral injection therapies or bulking agents have been variably effective in decreasing stress incontinence symptoms in patients with SUI, but they are routinely placed in the proximal urethral/bladder neck. An alternative technique would be to inject the muscular components of the external sphincter muscles and urethral crest to create an artificial seal of the sphincter complex without the need to surgically correct the support defect (**Figure 2(4)**). In patients with severe misalignment defects, injecting the EUS with enough of the agent to create a seal may not be possible without drawing back the anterior vaginal wall and posterior urethra by addressing apical prolapse. Understanding the role that voiding dysfunction, specifically strain voiding, plays in the generation of SUI is crucial to the success of this type of injection. Women with SUI generally do not relax their pelvic muscles to void and rely on the incompetency of the EUS to facilitate emptying. Enhancing the closure strength of the EUS could potentially cure SUI but also cause retention or LUTs in women who have not learned to relax their PR muscle and voluntarily open the EUS during voiding. Persistent post injection pushing behaviors could displace or dislodge a product before its full incorporation into the tissues, necessitating repeat treatment. Fortunately, women are easily incentivized to reverse their toileting habits when they are (1) alerted to the existence of the behaviors, and educated on their contribution to the leakage condition (2) told of the importance of correcting behaviors in order to have successful treatment outcomes and (3) taught how to voluntarily relax the PR muscle.

6. Correcting voiding dysfunction and CPMC

Most women with and without pelvic floor disorders would likely benefit from a tutorial on pelvic floor awareness and healthy toileting habits. Multiple professional societies have developed evidence-based guidelines for the evaluation and treatment of female UI, including the American Urological Association (AUA), the Society of Urodynamics, Female Pelvic Medicine and Urogenital Reconstruction (SUFU), the American College of Obstetricians and Gynecologists (ACOG), and the American Urogynecologic Society (AUGS), as well as international societies, such as the International Urogynecological Association (IUGA) [86, 87]. Behavioral management including dietary considerations, bladder retraining and pelvic floor exercises and/or pelvic floor therapy are the first line treatment of all pelvic floor disorders yet there is no standardized approach to these modalities nor are providers routinely trained on these techniques. Furthermore, referrals to outpatient pelvic floor therapists have poor attendance rates [88]. Ideally, the initial comprehensive visit should incorporate an introduction to pelvic floor awareness to ensure there are sensorimotor skills necessary for healthy toileting.

6.1 Manual therapy

During routine pelvic exam, the provider can guide the patient in her execution of three achievable pelvic floor positions and confirm her proprioceptive understanding.

The ability to relax and contract the muscle layers of the pelvic floor—the levator ani muscles and the PR muscle—is critical to the assessment of pelvic floor function [89]. Proper patient positioning on the exam table is crucial—supine recumbent with pressure off the buttocks onto the sacrum, and the legs bent with the soles of her feet flat on the table, is ideal for pelvic muscle assessment. The examiner gently inserts their finger into the vaginal opening and tests for sensation including discomfort. Normally, women will be in "clench" upon initial insertion. The highest/tightest position is "superclench" or Kegel, which is contraction of the lower abdominal muscles, the transversus abdominal, the gluteal muscles and the deep levator ani muscles to move the rectum and vagina forward against the bladder by elevating the levator plate [90, 91]. In this position, the superficial muscles, including the PR muscle, are also contracted (**Figure 3(1)**). Superclench involves core strength which is difficult to sustain due to muscle fatigue, but can be improved with core exercise therapy. The middle position is "clench" in which the deep muscles are relaxed but the PR and superficial muscles are contracted (**Figures 3(2)** and **4(1)**). Clench of the PR produces an anorectal angle which puts pressure on the openings of the urethra, vagina and anus to help store stool and urine and is the resting position for most women [92]. Both clench and superclench involve contraction of the PR but superclench adds the elevation of the levator plate. In order to urinate, defecate, or have comfortable sexual intercourse, women must release both the levator plate and the PR tension on these structures and move into the relaxed position (**Figures 3(3)** and **4(2)**). In general,

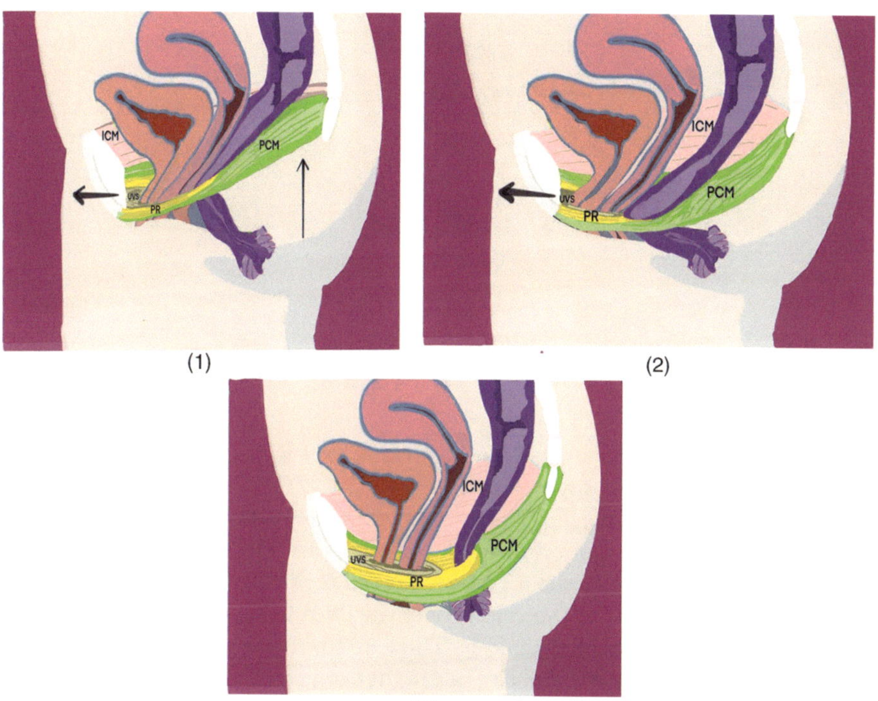

Figure 3.
Three positions of pelvic floor. (1) "Super clench". Elevated pubococcygeus (PCM), iliococcygeus (ICM), levator plate (LP). Contracted puborectalis (PR) and urethrovaginal sphincter (UVS). Kegel exercise. (2) "Clench". Relaxed PCM, ICM, levator plate. Contracted PR and UVS. Walk around position and (3) "relax". Relaxed PCM, ICM, LP. Relaxed PR and UVS. Position for micturition/elimination.

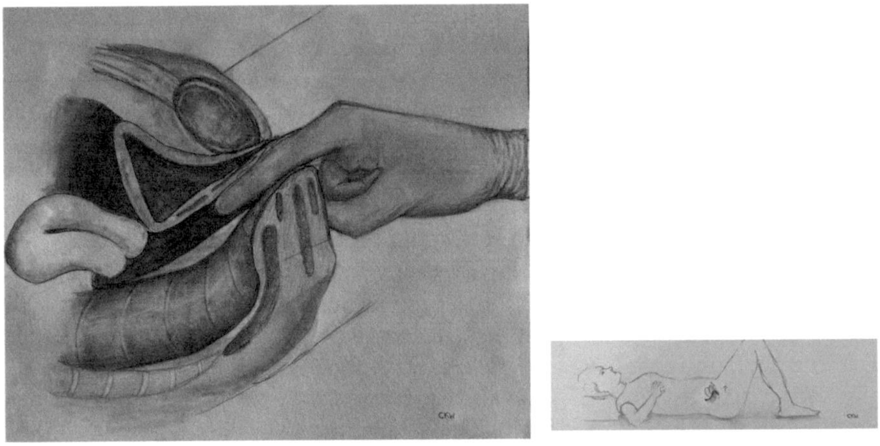

(1)

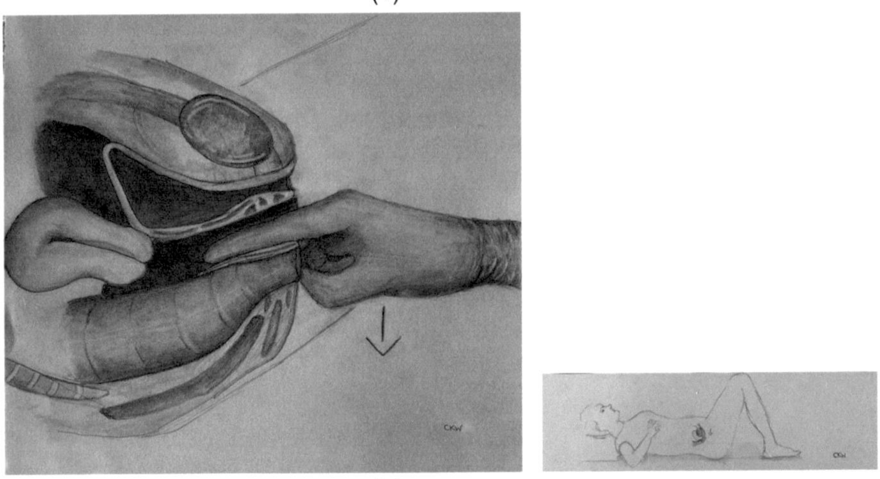

(2)

Figure 4.
Clench and relax—puborectalis release. (1) "Clench". Patient in semi supine with sacrum on the table. Puborectalis (PR) in raised position, (2) "relax". Position needed for elimination/micturition. Relaxed PR.

resting tone in the pelvic floor refers to the tone of the PR in clench, and strength refers to the grip strength of superclench.

Manual therapy serves the dual purpose of teaching proprioception and facilitating coordinated muscle motion through feedback provided by the examiner. Superclench is tested first and she is given feedback regarding holding strength. Gentle pressure on the back of the vagina over the PR muscle teaches her to voluntarily relax it, a technique described by Thiele in the 1960s [93]. Diaphragmatic breathing helps to relax the PR and can be taught even prior to exam. When the PR is released, the examiner tells her that she should be in this position during urination and defection (**Figure 1(1)**). Once the patient learns full relax, she is instructed to move from position to position gaining feedback from the examiner to confirm understanding. Myofascial puborectalis release has been shown to be effective in improving overactive bladder symptoms, and interstitial cystitis/PBS in multiple studies [94–96]. Adding a course of a pelvic floor relaxant such as vaginal Valium, or trigger point injections using various compounds including Botox have been shown to be effective in refractory cases. PRMT with or without biofeedback can reinforce learning how to operate the muscles in patients desiring more intensive training [97]. Biofeedback therapy has traditionally been used to treat dysfunctional voiding with good success in the pediatric population [98]. In a study of 958 postpartum women, instructional biofeedback was shown to have a beneficial effect in 73.6% [99]. It has also been shown to be useful in the correction of long term voiding dysfunction and recurrent urinary tract infections in adult women [100]. The most recent Cochrane review on the subject reported level 1 evidence that PFMT is successful in the conservative management of urinary incontinence [101].

6.2 Deep diaphragmatic breathing (DDP)

Normal breathing, also known as deep diaphragmatic breathing (DDP), involves synchronized motion of the upper rib cage, lower rib cage, and abdomen. Abnormal breathing, known as thoracic breathing, involves breathing from the upper chest, evidenced by greater upper rib cage motion, compared to the lower rib cage [102]. The pelvic floor muscles physiologically act as expiratory muscles in synergy with the anterolateral abdominal muscles, contracting during expiration and relaxing during inspiration. These coordinated movements facilitate blood flow and oxygen exchange without significant expenditure of energy during restful breathing [103]. When DDP is performed correctly, by inhaling through the nose, contracting the diaphragm and expanding the lower abdomen, there is an automatic release of the pelvic floor musculature, and external sphincters. Teaching patients to perform DDP while urinating and defecating is a useful modality to ensure complete pelvic floor and sphincter relaxation to facilitate unobstructed elimination. Patients can also be taught hypoopressive exercises (HE) which promote synergistic muscle movements of the diaphragm and abdominal wall utilizing DDP. One study showed that an 8-week HE program provided benefit in perineal function and decrease symptoms related to PFD and UI. In addition, HE had no adverse effects and treatment adherence was high (97%) [104].

6.3 Adult re-toilet training

Childhood toileting habits are deeply stored in primitive brain structures and unless they are pointed out and corrected, they are likely to persist. Women may not be cognizant of their voiding habits and should be asked to observe themselves to gain insights to their own patterns. Information on the negative effects of abdominal strain to void and defecate should be emphasized as this information is not widely available. Encouraging women to pay attention to and prioritize their own bodily needs is often the first step to developing bladder control. The main goal of bladder retraining is to learn to urinate without using an involuntary bladder contraction to void. In order to do so, women should monitor their fluid intake to be able to time her voiding episodes so she can void before the urge is strong [105]. She will train herself to relax the PR to initiate the bladder contraction and facilitate full emptying. Diaphragmatic breathing techniques can be used to voluntarily open the striated outlet muscles during voiding to allow the bladder to empty without pushing or straining. Depending on the degree of detrusor dysfunction, as bladder outlet obstruction is mindfully corrected, bladder stability can be restored. The process takes time and can be facilitated by overactive bladder medications if necessary. In cases where conservative methods have been unsuccessful, second line therapies such as Botox of the EUS/pelvic floor and neuromodulation, modalities used in pediatric patients with refractory BBD, can be considered [106, 107].

7. Conclusions

Unhealthy toileting habits are behavioral conditions which are established early in childhood and have the potential to manifest symptoms throughout the course of a woman's life. Depending on multiple factors, such as childbirth, hormonal status and psychosocial stress, the chronic failure to relax the pelvic outlet results in obstructed elimination and a wide spectrum of structural changes in the lower urinary tract anatomy which then become the focus of surgical treatment. Because these behaviors are unconscious, patient education to make her aware of the condition is integral to the process. In the absence of an underlying neurological condition, most women can reverse the dysfunctional toileting behaviors just by becoming informed of their existence. For some women, there is a need to have visceral feedback to understand the problem. These patients often demonstrate partial or complete sensorimotor dissociation from their pelvic floors muscles, unable to feel sensations or only to feel aberrant sensations and unable to exercise muscular control in this region. There is a need for validated and clear cut instructions on how to perform behavior therapies and exercises to help women learn to control the pelvic muscles, without the need for intensive pelvic floor therapy. On the toilet, deep diaphragmatic breathing techniques open the external sphincter to facilitate unobstructed voiding. Education on toileting habits by public health initiatives, pediatricians and gynecologists, prior to the onset of childbirth, will potentially avert structural changes later in life. Although pelvic floor awareness, education and re-toilet training may not benefit all cases of CPMC, motivated women can be taught to develop their interoceptive and pelvic motor skills, and observe their own toileting habits to correct their LUTD. This is especially important to undertake before any surgical procedure for urinary incontinence or prolapse is performed.

Conflict of interest

The author declares no conflict of interest. To all my teachers, most importantly the many patients who showed me what needed to be learned.

Notes/thanks/other declarations

C.G. Jung—'*Until you make the unconscious conscious*, it will direct your life and you will call it fate.'

Author details

Nicole Fleischmann
White Plains Hospital, White Plains, New York, United States

*Address all correspondence to: nfleischmann@gmail.com;
nfleischmann@wphospital.org

References

[1] Sjögren J, Malmberg L, Stenzelius K. Toileting behavior and urinary tract symptoms among younger women. International Urogynecology Journal. 2017;**11**:1677-1684. DOI: 10.1007/s00192-017-3319-2

[2] Daily AM, Kowalik CG, Delpe SD, Kaufman MR, Dmochowski RR, Reynolds WS. Women with overactive bladder exhibit more unhealthy toileting behaviors: A cross-sectional study. Urology. 2019;**134**:97-102. DOI: 10.1016/j.urology.2019.08.038

[3] Yang JM, Huang WC. Implications of abdominal straining in women with lower urinary tract symptoms. Urology. 2002;**60**:428-433. DOI: 10.1016/s0090-4295(02)01768-5

[4] Salinas J, Virseda M, Méndez S, Menéndez P, Esteban M, Moreno J. Abdominal strength in voiding cystometry: A risk factor for recurrent urinary tract infections in women. International Urogynecology Journal. 2015;**1**:1861-1865. DOI: 10.1007/s00192-015-2737-2

[5] Newman DK, Burgio KL, Cain C, Hebert-Beirne J, Low LK, Palmer MH, et al. Prevention of lower urinary tract symptoms (PLUS) research consortium. Toileting behaviors and lower urinary tract symptoms: A cross-sectional study of diverse women in the United States. International Journal of Nursing Studies Advances. 2021;**3**:100052. DOI: 10.1016/j.ijnsa.2021.100052

[6] Austin PF, Bauer SB, Bower W, Chase J, Franco I, Hoebeke P, et al. The standardization of terminology of lower urinary tract function in children and adolescents: Update report from the standardization committee of the International Children's Continence Society. Neurourology and Urodynamics. 2016;**35**(4):471-481. DOI: 10.1002/nau.22751

[7] Bower WF, Yip SK, Yeung CK. Dysfunctional elimination symptoms in childhood and adulthood. Journal of Urology. 2005;**174**(4 Pt 2):1623-1627; discussion 1627-8. Erratum in: J Urol. 2005;**174**(6):2428. DOI: 10.1097/01.ju.0000176599.91836.12

[8] Carlson KV, Rome S, Nitti VW. Dysfunctional voiding in women. The Journal of Urology. 2001;**165**(1):143-147; discussion 147-8. DOI: 10.1097/00005392-200101000-00035

[9] Fitzgerald MP, Thom DH, Wassel-Fyr C, Subak L, Brubaker L, Van Den Eeden SK, et al. Childhood urinary symptoms predict adult overactive bladder symptoms. The Journal of Urology. 2006;**175**(3 Pt 1):989-993. DOI: 10.1016/S0022-5347(05)00416-7

[10] Petrangeli F, Capitanucci ML, Marciano A, Mosiello G, Alvaro R, Zaccara A, et al. A 20-year study of persistence of lower urinary tract symptoms and urinary incontinence in young women treated in childhood. Journal of Pediatric Urology. 2014;**10**(3):441-445. DOI: 10.1016/j.jpurol.2014.01.003

[11] Minassian VA, Lovatsis D, Pascali D, Alarab M, Drutz HP. Effect of childhood dysfunctional voiding on urinary incontinence in adult women. Obstetrics and Gynecology. 2006;**107**(6):1247-1251. DOI: 10.1097/01.AOG.0000190222.12436.38

[12] Minassian VA, Langroudi MH, Parekh M, Poplawsky D, Lester

Kirchner H, Sartorius J. Childhood dysfunctional voiding is differentially associated with urinary incontinence subtypes in women. World Journal of Urology. 2012;**30**(1):111-115. DOI: 10.1007/s00345-011-0665-4

[13] Othman JA, Åkervall S, Molin M, Gyhagen M. Childhood nocturnal enuresis-a marker for pelvic floor disorders and urinary tract symptoms in women? International Urogynecology Journal. 2021;**32**(2):359-365. DOI: 10.1007/s00192-020-04345-x

[14] Akashi S, Tomita K. The impact of a history of childhood nocturnal enuresis on adult nocturia and urgency. Acta Paediatrica. 2014;**103**(9):e410-e415. DOI: 10.1111/apa.12694

[15] Goessaert AS, Schoenaers B, Opdenakker O, Hoebeke P, Everaert K, Vande Walle J. Long-term followup of children with nocturnal enuresis: Increased frequency of nocturia in adulthood. The Journal of Urology. 2014;**191**(6):1866-1870. DOI: 10.1016/j.juro.2013.12.051

[16] van Geen F-J, van de Wetering EHM, Nieuwhof-Leppink AJ, Klijn AJ, de Kort LMO. Dysfunctional voiding: Exploring disease transition from childhood to adulthood. Urology. 2023. DOI: 10.1016/j.urology.2023.03.018

[17] Pierce HM, Perry L, Gallagher R, Chiarelli P. Delaying voiding, limiting fluids, urinary symptoms, and work productivity: A survey of female nurses and midwives. Journal of Advanced Nursing. 2019;**75**(11):2579-2590. DOI: 10.1111/jan.14128

[18] Xu D, Zhu S, Li H, Gao J, Mou H, Wang K. Relationships among occupational stress, toileting behaviors, and overactive bladder in nurses: A multiple mediator model. Journal of Advanced Nursing. 2019;**75**(6):1263-1271. DOI: 10.1111/jan.13940

[19] Kowalik CG, Daily A, Delpe S, Kaufman MR, Fowke J, Dmochowski RR, et al. Toileting behaviors of women-what is healthy? The Journal of Urology. 2019;**201**(1):129-134. DOI: 10.1016/j.juro.2018.07.044

[20] Iyer J, Rane A. Effects of posture and squatting on the dynamics of micturition: Response to Bush and Liedl. International Urogynecology Journal. 2015;**26**(5):781. DOI: 10.1007/s00192-015-2646-4

[21] Pauwels E, De Wachter S, Wyndaele JJ. A normal flow pattern in women does not exclude voiding pathology. International Urogynecology Journal and Pelvic Floor Dysfunction. 2005;**16**(2):104-108; discussion 108. DOI: 10.1007/s00192-004-1227-8

[22] Neels H, Wyndaele JJ, Tjalma WA, De Wachter S, Wyndaele M, Vermandel A. Knowledge of the pelvic floor in nulliparous women. Journal of Physical Therapy Science. 2016;**28**(5):1524-1533. DOI: 10.1589/jpts.28.1524

[23] El-Hamamsy D, Parmar C, Shoop-Worrall S, Reid FM. Public understanding of female genital anatomy and pelvic organ prolapse (POP); a questionnaire-based pilot study. International Urogynecology Journal. 2022;**33**(2):309-318. DOI: 10.1007/s00192-021-04727-9

[24] Michael J, McFarland J. Another look at the core concepts of physiology: Revisions and resources. Advances in Physiology Education. 2020;**44**:752-762

[25] Kose O, Atik YT, Gul D, Uysal B, Cimen HI, Bostanci MS. Influence of toileting behavior on the natural course

of anterior vaginal wall prolapse. BMC Women's Health. 2022;**22**(1):56. DOI: 10.1186/s12905-022-01637-w

[26] de Groat WC, Griffiths D, Yoshimura N. Neural control of the lower urinary tract. Comprehensive Physiology. 2015;**5**(1):327-396. DOI: 10.1002/cphy.c130056. Study. Front Hum Neurosci. 2022;**16**:912839. DOI: 10.3389/fnhum.2022.912839

[27] Sillén U. Bladder function in healthy neonates and its development during infancy. The Journal of Urology. 2001;**166**(6):2376-2381. DOI: 10.1016/s0022-5347(05)65594-2

[28] Jansson UB, Hanson M, Sillén U, Hellström AL. Voiding pattern and acquisition of bladder control from birth to age 6 years—a longitudinal study. The Journal of Urology. 2005;**174**(1):289-293. DOI: 10.1097/01.ju.0000161216.45653.e3

[29] Nereus T, Sillén U. Lower urinary tract function in childhood; normal development and common functional disturbances. Acta Physiologica (Oxford, England). 2013;**207**(1):85-92. DOI: 10.1111/apha.12015

[30] Vermandel A, Van Kampen M, Van Gorp C, Wyndaele JJ. How to toilet train healthy children? A review of the literature. Neurourology and Urodynamics. 2008;**27**(3):162-166. DOI: 10.1002/nau.20490

[31] Bloom DA, Park JM, Koo HP. Comments on pediatric elimination dysfunctions: The Whorf hypothesis, the elimination interview, the guarding reflex and nocturnal enuresis. European Urology. 1998;**33**(Suppl. 3):20-24. DOI: 10.1159/000052237

[32] Nevéus T, von Gontard A, Hoebeke P, Hjälmås K, Bauer S, Bower W, et al. The standardization of terminology of lower urinary tract function in children and adolescents: Report from the standardisation Committee of the International Children's continence society. The Journal of Urology. 2006;**176**(1):314-324. DOI: 10.1016/S0022-5347(06)00305-3

[33] Sureshkumar P, Jones M, Cumming R, Craig J. A population based study of 2,856 school-age children with urinary incontinence. The Journal of Urology. 2009;**181**(2):808-815; discussion 815-6. DOI: 10.1016/j.juro.2008.10.044

[34] Hinman F Jr. Nonneurogenic neurogenic bladder (the Hinman syndrome)--15 years later. The Journal of Urology. 1986;**136**(4):769-777. DOI: 10.1016/s0022-5347(17)45077-4

[35] Osman NI, Chapple CR. Fowler's syndrome—A cause of unexplained urinary retention in young women? Nature Reviews. Urology. 2014;**11**(2):87-98. DOI: 10.1038/nrurol.2013.277

[36] Hoeritzauer I, Stone J, Fowler C, Elneil-Coker S, Carson A, Panicker J. Fowler's syndrome of urinary retention: A retrospective study of co-morbidity. Neurourology and Urodynamics. 2016;**35**:601-603

[37] vonGontard A, Niemczyk J, Weber M, Equit M. Specific behavioral comorbidity in a large sample of children with functional incontinence: Report of 1,001 cases. Neurourology and Urodynamics. 2015;**34**:763-768

[38] Brady SS, Arguedas A, Huling JD, Shan L, Lewis CE, Fok CS, et al. Adverse childhood experiences and lower urinary tract symptoms and impact among women. The Journal of Urology. 2023;**209**(6):1167-1175. DOI: 10.1097/JU.0000000000003387

[39] Dourado ER, de Abreu GE, Santana JC, Macedo RR, da Silva CM,

Rapozo PMB, et al. Emotional and behavioral problems in children and adolescents with lower urinary tract dysfunction: A population-based study. Journal of Pediatric Urology. 2019;**15**(4):376.e1-376.e7. DOI: 10.1016/j.jpurol.2018.12.003

[40] von Gontard A, Baeyens D, Van Hoecke E, et al. Psychological and psychiatric issues in urinary and fecal incontinence. The Journal of Urology. 2011;**185**:1432-1437

[41] Logan BA, Correia K, McCarthy J, Slattery MJ. Voiding dysfunction related to adverse childhood experiences and neuropsychiatric disorders. Journal of Pediatric Urology. 2014;**10**(4):634-638. DOI: 10.1016/j.jpurol.2014.06.012

[42] Lai HH, Rawal A, Shen B, Vetter J. The relationship between anxiety and overactive bladder or urinary incontinence symptoms in the clinical population. Urology. 2016;**98**:50-57. DOI: 10.1016/j.urology.2016.07.013

[43] Wood SK, Baez MA, Bhatnagar S, Valentino RJ. Social stress-induced bladder dysfunction: Potential role of corticotropin-releasing factor. American Journal of Physiology. Regulatory, Integrative and Comparative Physiology. 2009;**296**(5):R1671-R1678. DOI: 10.1152/ajpregu.91013.2008

[44] Joinson C, Heron J, Butler U, von Gontard A. Psychological differences between children with and without soiling problems. Pediatrics. 2006;**117**(5):1575-1584

[45] Melotti IGR, Juliato CRT, Tanaka M, Riccetto CLZ. Severe depression and anxiety in women with overactive bladder. Neurourology and Urodynamics. 2018;**37**(1):223-228. DOI: 10.1002/nau.23277

[46] Chapel CR, Osman NI, Birder L, Dmochowski R, Drake MJ, van Koeveringe G, et al. Terminology report from the international continence society (ICS) Working Group on Underactive Bladder (UAB). Neurourology and Urodynamics. 2018;**37**(8):2928-2931. DOI: 10.1002/nau.23701

[47] Valentini FA, Marti BG, Robain G. Do urodynamics provide a better understanding of voiding disorders in women over 80? Progrès en Urologie. 2018;**28**(4):230-235

[48] Pauwels E, De Laet K, Der Wachter S, Wyndaele JJ. Healthy, middle-aged, history-free, continent women--do they strain to void? Journal of Urology. 2006;**175**:1403-1407

[49] Nitti VW, Tu LM, Gitlin J. Diagnosing bladder outlet obstruction in women. The Journal of Urology. 1999;**161**(5):1535-1540

[50] Everaert K, Van Laecke E, De Muynck M, Peeters H, Hoebeke P. Urodynamic assessment of voiding dysfunction and dysfunctional voiding in girls and women. International Urogynecology Journal and Pelvic Floor Dysfunction. 2000;**11**(4):254-264. DOI: 10.1007/s001920070035

[51] Devreese AM, Nuyens G, Staes F, Vereecken RL, De Weerdt W, Stappaerts K. Do posture and straining influence urinary-flow parameters in normal women? Neurourology and Urodynamics. 2000;**19**:3-8

[52] Gammie A, Kaper M, Dorrepaal C, Kos T, Abrams P. Signs and symptoms of detrusor underactivity: An analysis of clinical presentation and urodynamic tests from a large group of patients undergoing pressure flow studies. European Urology. 2016;**69**(2):361-369

[53] Lemack GE, Baseman AG, Zimmern PE. Voiding dynamics in women: A comparison of pressure-flow studies between asymptomatic and incontinent women. Urology. 2002;**59**(1):42-46. DOI: 10.1016/s0090- 4295(01)01462-5

[54] Karram MM, Partoll L, Bilotta V, Angel O. Factors affecting detrusor contraction strength during voiding in women. Obstetrics and Gynecology. 1997;**90**(5):723-726. DOI: 10.1016/S0029-7844(97)00439-0

[55] Uren AD, Drake MJ. Definition and symptoms of underactive bladder. Investigative and Clinical Urology. 2017;**58**(Suppl. 2):S61-S67. DOI: 10.4111/icu.2017.58.S2.S61. Epub 2017 Nov 13

[56] Miller EA, Amundsen CL, Toh KL, Flynn BJ, Webster GD. Preoperative urodynamic evaluation may predict voiding dysfunction in women undergoing pubovaginal sling. The Journal of Urology. 2003;**169**(6):2234-2237. DOI: 10.1097/01.ju.0000063590.13100.4d

[57] Iglesia CB, Shott S, Fenner DE, Brubaker L. Effect of preoperative voiding mechanism on success rate of autologous rectus fascia suburethral sling procedure. Obstetrics and Gynecology. 1998;**91**(4):577-581. DOI: 10.1016/s0029-7844(98)00029-5

[58] Chen L, Ashton-Miller JA, DeLancey JOL. A 3D finite element model of anterior vaginal wall support to evaluate mechanisms underlying cystocele formation. Journal of Biomechanics. 2009;**42**(10):1371-1377. DOI: 10.1016/j.jbiomech.2009.04.043

[59] Frigerio M, Manodoro S, Cola A, Palmieri S, Spelzini F, Milani R. Detrusor underactivity in pelvic organ prolapse. International Urogynecology Journal. 2018;**29**(8):1111-1116. DOI: 10.1007/s00192-017-3532-z. Epub 2017 Dec 21

[60] Kahn MA, Breitkopf CR, Valley MT, Woodman PJ, O'Boyle AL, Bland DI, et al. Pelvic Organ Support Study (POSST) and bowel symptoms: Straining at stool is associated with perineal and anterior vaginal descent in a general gynecologic population. American Journal of Obstetrics and Gynecology. 2005;**192**(5):1516-1522. DOI: 10.1016/j.ajog.2004.11.040

[61] Seki N, Shahab N, Hara R, Takei M, Yamaguchi A, Naito S. Voiding dynamics in women with stress urinary incontinence and high-stage cystocele. International Journal of Urology. 2011;**18**(3):219-224. DOI: 10.1111/j.1442-2042.2010.02697.x

[62] Herschorn S. Female pelvic floor anatomy: The pelvic floor, supporting structures, and pelvic organs. Revista de Urología. 2004;**6**(Suppl. 5):S2-S10

[63] Azpiroz F, Fernandez-Fraga X, Merletti R, Enck P. The puborectalis muscle. Neurogastroenterology and Motility. 2005;**17**(Suppl. 1):68-72. DOI: 10.1111/j.1365-2982.2005.00663.x

[64] Raizada V, Bhargava V, Jung SA, Karstens A, Pretorius D, Krysl P, et al. Dynamic assessment of the vaginal high-pressure zone using high-definition manometery, 3-dimensional ultrasound, and magnetic resonance imaging of the pelvic floor muscles. American Journal of Obstetrics and Gynecology. 2010;**203**(2):172.e1-172.e8. DOI: 10.1016/j.ajog.2010.02.028

[65] Williams AMM, Sato-Klemm M, Deegan EG, Eginyan G, Lam T. Characterizing pelvic floor muscle activity during walking and jogging in continent adults:

A cross-sectional study. Frontiers in Human Neuroscience. 2022;**16**:912839. DOI: 10.3389/fnhum.2022.912839

[66] Frawley H, Shelly B, Morin M, Bernard S, Bø K, Digesu GA, et al. An International Continence Society (ICS) report on the terminology for pelvic floor muscle assessment. Neurourology and Urodynamics. 2021;**40**(5):1217-1260. DOI: 10.1002/nau.24658

[67] Peters KM, Carrico DJ, Kalinowski SE, Ibrahim IA, Diokno AC. Prevalence of pelvic floor dysfunction in patients with interstitial cystitis. Urology. 2007;**70**(1):16-18. DOI: 10.1016/j.urology.2007.02.067

[68] Kadah S, Soh SE, Morin M, Schneider M, Heron E, Frawley H. Is there a difference in pelvic floor muscle tone between women with and without pelvic pain? A systematic review and meta-analysis. The Journal of Sexual Medicine. 2023;**20**(1):65-96. DOI: 10.1093/jsxmed/qdac002

[69] Fan YH, Lin AT, Wu HM, Hong CJ, Chen KK. Psychological profile of female patients with dysfunctional voiding. Urology. 2008;**71**:625-629

[70] Vrijens D, Berghmans B, Nieman F, van Os J, van Koeveringe G, Leue C. Prevalence of anxiety and depressive symptoms and their association with pelvic floor dysfunctions—A cross sectional cohort study at a Pelvic Care Centre. Neurourology and Urodynamics. 2017;**36**(7):1816-1823. DOI: 10.1002/nau.23186

[71] Neels H, Tjalma WA, Wyndaele JJ, De Wachter S, Wyndaele M, Vermandel A. Knowledge of the pelvic floor in menopausal women and in peripartum women. Journal of Physical Therapy Science. 2016;**28**(11):3020-3029

[72] Fitz FF, Paladini LM, Ferreira LA, Gimenez MM, Bortolini MAT, Castro RA. Ability to contract the pelvic floor muscles and association with muscle function in incontinent women. International Urogynecology Journal. 2020;**31**(11):2337-2344. DOI: 10.1007/s00192-020-04469-0

[73] Miotto VG, Fernades ACNL, de la Ossa AMP, Bazanelli GS, Fretta TB, de Paula NA, et al. Women with functional constipation have a worse ability to relax their pelvic floor muscle: A cross sectional study. Neurourology and Urodynamics. 2023;**42**(1):123-132. DOI: 10.1002/nau.25057

[74] Chiarelli P, Murphy B, Cockburn J. Women's knowledge, practises, and intentions regarding correct pelvic floor exercises. Neurourology and Urodynamics. 2003;**22**(3):246-249. DOI: 10.1002/nau.10119

[75] Hanno PM, Erickson D, Moldwin R, Faraday MM, American Urological Association. Diagnosis and treatment of interstitial cystitis/bladder pain syndrome: AUA guideline amendment. The Journal of Urology. 2015;**193**(5):1545-1553. DOI: 10.1016/j.juro.2015.01.086

[76] Pisani GK, de Oliveira ST, Carvalho C. Pelvic floor dysfunctions and associated factors in female CrossFit practitioners: A cross-sectional study. International Urogynecology Journal. 2021;**32**(11):2975-2984. DOI: 10.1007/s00192-020-04581-1

[77] Dietz HP, Shek KL. The quantification of levator muscle resting tone by digital assessment. International Urogynecology Journal and Pelvic Floor Dysfunction. 2008;**19**(11):1489-1493. DOI: 10.1007/s00192-008-0682-z

[78] Hundley AF, Wu JM, Visco AG. A comparison of perineometer to

brink score for assessment of pelvic floor muscle strength. American Journal of Obstetrics and Gynecology. 2005;**192**(5):1583-1591. DOI: 10.1016/j.ajog.2004.11.015

[79] Thompson JA, O'Sullivan PB, Briffa NK, Neumann P. Assessment of voluntary pelvic floor muscle contraction in continent and incontinent women using transperineal ultrasound, manual muscle testing and vaginal squeeze pressure measurements. International Urogynecology Journal and Pelvic Floor Dysfunction. 2006;**17**:624-630

[80] van Os-Bossagh P, Pols T, Hop WC, et al. Voiding symptoms in chronic pelvic pain (CPP). European Journal of Obstetrics, Gynecology, and Reproductive Biology. 2003;**107**(2):185-190

[81] Ashton-Miller JA, DeLancey JO. Functional anatomy of the female pelvic floor. Annals of the New York Academy of Sciences. 2007;**1101**:266-296. DOI: 10.1196/annals.1389.034

[82] Deindl FM, Vodusek DB, Bischoff CH, et al. Dysfunctional voiding in women: Which muscles are responsible? British Journal of Urology. 1998;**82**:814-819

[83] Delancey JO, Ashton-Miller JA. Pathophysiology of adult urinary incontinence. Gastroenterology. 2004;**126**(Suppl. 1):S23-S32

[84] Petros PE, Ulmsten UI. An integral theory and its method for the diagnosis and management of female urinary incontinence. Scandinavian Journal of Urology and Nephrology. Supplementum. 1993;**153**:1-93

[85] Ludwig S, Göktepe S, Mallmann P, Jäger W. Evaluation of different 'Tensioning' of apical suspension in women undergoing surgery for prolapse and urinary incontinence. In Vivo. 2020;**34**(3):1371-1375

[86] Haylen BT, de Ridder D, Freeman RM, et al. An International Urogynecological Association (IUGA)/International Continence Society (ICS) joint report on the terminology for female pelvic floor dysfunction. Neurourology and Urodynamics. 2010;**29**(1):4-20. DOI: 10.1002/nau.20798

[87] Sussman RD, Syan R, Brucker BM. Guideline of guidelines: Urinary incontinence in women. BJU International. 2020;**125**(5):638-655. DOI: 10.1111/bju.14927

[88] Pan LC, Datar M, McKinney JL, Keyser LE, Goss TF, Pulliam SJ. Adherence to professional society guidelines among women with stress or mixed urinary incontinence. Neurourology and Urodynamics. 2022;**41**(6):1489-1497. DOI: 10.1002/nau.24986

[89] Hoebeke P, Vande Walle J, Theunis M, De Paepe H, Oosterlinck W, Renson C. Out-patient pelvic floor therapy in girls with daytime incontinence of dysfunctional voiding. Urology. 1996;**48**:923-927

[90] Park H, Han D. The effect of the correlation between the contraction of the pelvic floor muscles and diaphragmatic motion during breathing. Journal of Physical Therapy Science. 2015;**27**(7):2113-2115. DOI: 10.1589/jpts.27.2113

[91] Kegel AH. Progressive resistance exercise in the functional restoration of the perineal muscles. American Journal of Obstetrics and Gynecology. 1948;**56**(2):238-248. DOI: 10.1016/0002-9378(48)90266-x

[92] Neumann P, Gill V. Pelvic floor and abdominal muscle interaction: EMG activity and intra-abdominal pressure. International Urogynecology Journal. 2002;**13**:125-132

[93] Thiele GH. Coccygodynia: Cause and treatment. Diseases of the Colon and Rectum. 1963;**6**:422-436

[94] Weiss JM. Pelvic floor myofascial trigger points: Manual therapy for interstitial cystitis and the urgency-frequency syndrome. The Journal of Urology. 2001;**166**(6):2226-2231. DOI: 10.1016/s0022-5347(05)65539-5

[95] Wolff BJ, Joyce CJ, McAlarnen LA, Brincat CA, Mueller ER, Fitzgerald CM. Consideration of pelvic floor myofascial release for overactive bladder. Journal of Bodywork and Movement Therapies. 2020;**24**(2):144-150. DOI: 10.1016/j.jbmt.2019.10.018

[96] Oyama IA, Rejba A, Lukban JC, et al. Modified Thiele massage as therapeutic intervention for female patients with interstitial cystitis and high-tone pelvic floor dysfunction. Urology. 2004;**64**:862-865

[97] Clemens JQ, Nadler RB, Schaeffer AJ, et al. Biofeedback, pelvic floor re-education, and bladder training for male chronic pelvic pain syndrome. Urology. 2000;**56**:951-955

[98] Porena M, Costantini E, Rociola W, Mearini E. Biofeedback successfully cures detrusor-sphincter dyssynergia in pediatric patients. The Journal of Urology. 2000;**163**(6):1927-1931

[99] Vermandel A, De Wachter S, Beyltjens T, et al. Pelvic floor awareness and the positive effect of verbal instructions in 958 women early postdelivery. International Urogynecology Journal. 2015;**26**:223-228

[100] Minardi D, d'Anzeo G, Parri G, Polito M Jr, Piergallina M, El Asmar Z, et al. The role of uroflowmetry biofeedback and biofeedback training of the pelvic floor muscles in the treatment of recurrent urinary tract infections in women with dysfunctional voiding: A randomized controlled prospective study. Urology. 2010;**75**(6):1299-1304. DOI: 10.1016/j.urology.2009.11.019

[101] Todhunter-Brown A, Hazelton C, Campbell P, Elders A, Hagen S, McClurg D. Conservative interventions for treating urinary incontinence in women: An overview of Cochrane systematic reviews. Cochrane Database of Systematic Reviews. 2022;**9**(9):CD012337. DOI: 10.1002/14651858.CD012337.pub2

[102] Barker NJ, Jones M, O'Connell NE, Everard ML. Breathing exercises for dysfunctional breathing/hyperventilation syndrome in children. Cochrane Database of Systematic Reviews. 2013;(12):CD010376. DOI: 10.1002/14651858.CD010376.pub2

[103] Talasz H, Kremser C, Talasz HJ, Kofler M, Rudisch A. Breathing, (S) training and the pelvic floor-a basic concept. Healthcare (Basel). 2022;**10**(6):1035. DOI: 10.3390/healthcare10061035

[104] Molina-Torres G, Moreno-Muñoz M, Rebullido TR, Castellote-Caballero Y, Bergamin M, Gobbo S, et al. The effects of an 8-week hypopressive exercise training program on urinary incontinence and pelvic floor muscle activation: A randomized controlled trial. Neurourology and Urodynamics. 2023;**42**(2):500-509. DOI: 10.1002/nau.25110

[105] Ostaszkiewicz J, Johnston L, Roe B. Habit retraining for the management of urinary incontinence in adults.

Cochrane Database of Systematic Reviews. 2004;**2004**(2):CD002801. DOI: 10.1002/14651858.CD002801.pub2

[106] Radojicic ZI, Perovic SV, Milic NM. Is it reasonable to treat refractory voiding dysfunction in children with botulinum-a toxin? The Journal of Urology. 2006;**176**(1):332-336; discussion 336. DOI: 10.1016/S0022-5347(06)00298-9

[107] Dwyer ME, Vandersteen DR, Hollatz P, Reinberg YE. Sacral neuromodulation for the dysfunctional elimination syndrome: A 10-year single-center experience with 105 consecutive children. Urology. 2014;**84**(4):911-917. DOI: 10.1016/j.urology.2014.03.059